HENRY P.G. DARCY AND OTHER PIONEERS IN HYDRAULICS

CONTRIBUTIONS IN CELEBRATION OF THE 200TH BIRTHDAY
OF HENRY PHILIBERT GASPARD DARCY

June 23–26, 2003
Philadelphia, PA

SPONSORED BY
Environmental and Water Resources Institute (EWRI)
of the American Society of Civil Engineers

Surface Water Hydrology Committee
History and Heritage Committee

EDITED BY
Glenn O. Brown
Jürgen D. Garbrecht
Willi H. Hager

Published by the American Society of Civil Engineers

Library of Congress Cataloging-in-Publication Data

Henry P.G. Darcy and other pioneers in hydraulics : contributions in celebration of the 200th birthday of Henry Philibert Gaspard Darcy, June 23–26, 2003, Philadelphia, PA / sponsored by Environmental and Water Resources Institute of the American Society of Civil Engineers, Surface Water Hydrology Committee, History and Heritage Committee ; edited by Glenn O. Brown, Jürgen D. Garbrecht, Willi H. Hager.
  p. cm.
Collection of contributions, most of which will be presented at the Darcy Memorial Symposium on the History of Hydraulics on June 24, 2003, in Philadelphia, Pennsylvania.
  ISBN 0-7844-0683-9
  1. Hydraulics--History--Congresses. 2. Darcy, Henry, 1803-1858--Congresses. I. Darcy, Henry, 1803-1858. II. Brown, Glenn O. (Glenn Owen) III. Garbrecht, Jürgen. IV. Hager, Willi H. V. Environmental and Water Resources Institute (U.S.) VI. Darcy Memorial Symposium on the History of Hydraulics (2003 : Philadelphia, Pa.)

TC160.H46 2003
627'.09--dc21

2003050258

Published by American Society of Civil Engineers
1801 Alexander Bell Drive
Reston, Virginia 20191
www.asce pubs.asce.org

Any statements expressed in these materials are those of the individual authors and do not necessarily represent the views of ASCE, which takes no responsibility for any statement made herein. No reference made in this publication to any specific method, product, process or service constitutes or implies an endorsement, recommendation, or warranty thereof by ASCE. The materials are for general information only and do not represent a standard of ASCE, nor are they intended as a reference in purchase specifications, contracts, regulations, statutes, or any other legal document. ASCE makes no representation or warranty of any kind, whether express or implied, concerning the accuracy, completeness, suitability, or utility of any information, apparatus, product, or process discussed in this publication, and assumes no liability therefore. This information should not be used without first securing competent advice with respect to its suitability for any general or specific application. Anyone utilizing this information assumes all liability arising from such use, including but not limited to infringement of any patent or patents.

ASCE and American Society of Civil Engineers—Registered in U.S. Patent and Trademark Office.

Photocopies: Authorization to photocopy material for internal or personal use under circumstances not falling within the fair use provisions of the Copyright Act is granted by ASCE to libraries and other users registered with the Copyright Clearance Center (CCC) Transactional Reporting Service, provided that the base fee of $18.00 per article is paid directly to CCC, 222 Rosewood Drive, Danvers, MA 01923. The identification for ASCE Books is 0-7844-0683-9/03/ $18.00. Requests for special permission or bulk copying should be addressed to Permissions & Copyright Dept., ASCE.

Copyright © 2003 by the American Society of Civil Engineers.
All Rights Reserved.
Library of Congress Catalog Card No: 2003050258
ISBN 0-7844-0683-9
Manufactured in the United States of America.

HENRY PHILIBERT GASPARD DARCY
Circa 1856
From the collection of Jean Darcy, Paris.

Abstract

June 10th, 2003 is the 200th anniversary of the birth of Henry Philibert Gaspard Darcy (1803-1858). While best known for his research findings of Darcy's Law and the Darcy-Weisbach Equation, he was also a skilled engineer, public servant, and community leader. In celebration of the anniversary, this publication is a collection of 20 peer-reviewed contributions that address both Darcy and other pioneers in hydraulic history. Seven papers describe Darcy's achievements and legacy, and include a paper contributed by Darcy's descendent and namesake. Additional essays address the contributions of Henry Bazin, Günther Garbrecht, Hans Einstein, Victor Streeter, Carl Kindsvater, and Floyd Nagler. Historical developments in pumping equipment in medieval Europe and the USA in the 19th century, dam failures due to uplift, and the hydraulic engineering collection at the Smithsonian Institution are also reviewed. Finally, accounts for the Colorado State University, St. Anthony Falls, and USDA-ARS laboratories are presented.

Résumé

Le 10 juin, 2003 est le bicentenaire de la naissance d'Henry Philibert Gaspard Darcy (1803-1858). Tandis que connu pour ses résultats de recherches sur la loi de Darcy et sur l'équation de Darcy-Weisbach, il était un fonctionnaire, et un chef de la communauté, et également un ingénieur habile. Dans la célébration de l'anniversaire, cette publication est une collection de 20 contributions revues par des pairs qui adressent Darcy et d'autres pionniers dans l'histoire hydraulique. Sept articles décrivent les accomplissements et le legs de Darcy, dont un article contribué par le descendant de Darcy. Les essais additionnels adressent les contributions d'Henry Bazin, Günther Garbrecht, Hans Einstein, Victor Streeter, Carl Kindsvater, et Floyd Nagler. Sont également passés en revue : les développements historiques dans l'équipement de pompage en à l'Europe médiévale et aux Etats-Unis dans le 19ème siècle, les échecs de barrage dus au soulèvement, et la collection du génie hydraulique à l'Institution Smithsonian. En conclusion, les comptes pour les laboratoires de l'Université de l'Etat du Colorado, de St. Anthony Falls, et de USDA-ARS, sont présentés.

Foreword

June 10th, 2003 marks the 200th anniversary of the birth of Henry Philibert Gaspard Darcy (1803-1858). Darcy was a person of unusual ability and accomplishment. He is credited with inventing the modern style Pitot tube, was the first researcher to suspect the existence of the boundary layer in fluid flow, contributed in the development of the Darcy-Weisbach equation for pipe flow resistance, made major contributions to open channel flow research, provided the first quantitative measurements of artesian well flow, and of course, developed Darcy's Law for flow in porous media. His work provides foundations for several fields of study including hydraulics, ground-water hydrology, soil physics, and petroleum engineering. In addition to his research accomplishments, he was a skilled engineer, public servant, and community leader.

In celebration of the anniversary and to draw appropriate recognition to Darcy's work, the Environmental and Water Resources Institute of the American Society of Civil Engineers will hold the *Darcy Memorial Symposium on the History of Hydraulics* on June 24, 2003. The Symposium is part of the larger World Water Congress to be held in Philadelphia, Pennsylvania. The specialty symposium commemorates both Darcy's accomplishments and other more recent pioneers in hydraulic research and practice. This publication is a collection of contributions, most of which will be presented at the symposium. Contributions were only accepted after meeting editorial standards and passing an anonymous peer review.

Starting in 2001, the Task Committee, *Darcy Memorial Congress on the History of Hydraulics*, was sponsored by the Surface Water Hydrology Committee and was later joined by the new History and Heritage Committee of EWRI. The editors wish to thank the American Society of Civil Engineers, the ASCE publication staff, reviewers and contributors who helped in the production of this book. Jerry Rogers of the University of Houston, who served as a Co-Chair of the task committee and Chair of History and Heritage Committee, provided important guidance to the planning effort. Finally, Elaina Wright, Unit Assistant, Oklahoma State University preformed many of the editorial tasks and ensured a timely completion of this work.

The EWRI and the History and Heritage Committee are committed to regularly publish books such as this one. Currently in production is "*History of Hydraulic Equations*", which is intended for use in teaching. Jerry Rogers is also organizing the *Water Resources and Environmental Symposium* to be held at the 2004 Salt Lake City, Utah, EWRI Congress. These activities present our hope for the continued documentation of the history of the water resources and environmental profession.

-- Glenn O. Brown, Jürgen D. Garbrecht and Willi H. Hager, editors

Contents

Note: Spell it "Henry Darcy" 1
   Glenn O. Brown and Willi H. Hager

Henry Darcy's Contributions

Henry Darcy: Inspecteur général des ponts et chausses 4
   Henry Darcy

Henry Darcy's Perfection of the Pitot Tube 14
   Glenn O. Brown

Henry Darcy and the Pipe Flow Formula 24
   Corrado Gisonni

Henry Darcy and the Public Fountains of the City of Dijon 37
   Patricia Bobeck

Henry Darcy: Biography by Caudemberg 51
   Willi H. Hager

Henry Darcy's Legacy

The Place of Darcy's Law in the Framework of Non-Equilibrium Thermodynamics 71
   P.H. Groenevelt

Darcy's Law from Water to the Petroleum Industry: When and Who? 78
   Paolo Macini and Ezio Mesini

Henry Bazin: Hydraulician 90
   Willi H. Hager and Corrado Gisonni

Hydraulic Pioneers

Water Management and Hydraulic Structures in Antiquity:
Contributions by Günther Garbrecht 116
   Jürgen Garbrecht and Henning Fahlbusch

Hans Albert Einstein's Efforts to Understand and Formulate Bed-Sediment
Transport in Rivers 140
   Robert Ettema and Cornelia F. Mutel

A Tribute to Victor L. Streeter 160
    D.C. Wiggert and E.B. Wylie

A Tribute to Carl Kindsvater and the Georgia Tech Hydraulics Laboratory 174
    Terry W. Sturm and C. Samuel Martin

Floyd Nagler, Founding Director of IIHR 199
    Cornelia F. Mutel and Robert Ettema

Hydraulic Practice in History

Historical Improvements in Groundwater-Pumping Equipment
and Effects on Farming in the United States 207
    Bob Kent and Greg Hamer

It is a Crime to Design a Dam Without Considering Upward Pressure:
Engineers and Uplift, 1890-1930 220
    Donald C. Jackson

The Evolution of Pumping Systems Through the Early Renaissance 233
    Paolo Macini and Ezio Mesini

Hydraulic Engineering History at the Smithsonian Institution 252
    William E. Worthington, Jr.

Hydraulic Laboratories

A Historic Look at the USDA-ARS Hydraulic Engineering Research Unit 263
    Sherry L. Britton, Gregory J. Hanson, and Darrel M. Temple

History of Hydraulics and Fluid Mechanics at Colorado State University 277
    Pierre Y. Julien and Robert N. Meroney

The St. Anthony Falls Laboratory in History 289
    Ed Silberman, Roger Arndt, Gary Parker, Efi Foufoula-Georgiou,
    and Chris Paola

Subject Index 309

Author Index 311

Note: Spell it "Henry Darcy"

Glenn O. Brown[1] and Willi H. Hager[2]

Abstract

Henry Darcy's first name is commonly spelled Henri, while his last name sometimes appears as d'Arcy. Original source material shows that the correct spelling is, "Henry Darcy". The "Henry" spelling was his from birth, while the "Darcy" spelling was adopted in his youth and kept throughout his life.

Issues on the Spelling

At the start of this volume, it may be appropriate to comment on Henry Philibert Gaspard Darcy's name. He wrote his first name as "Henry", not in the French fashion "Henri". His preferred spelling may be seen in the title pages of his most important publications *Les fontaines publiques de la ville de Dijon* (Darcy 1856) *and Recherches expérimentales relatives au mouvement de l'eau dans les tuyaux* (Darcy 1857), and also on his grave (Philip 1995). Recently, the second author has found his birth certificate in Dijon. As can be see in Figure 1, the first name was spelled "Henry". Note also that the date of birth is given in the Revolutionary Calendar, not as we note it, June 10, 1803.

The confusion about his given name arises from two factors. First, the practice of his day was to neglect first names or use only initials. His first known publication is signed simply "L'ingénieur ordinaire, Darcy" (Darcy 1834), while his two most easily found publications list the author as "H. Darcy", (Darcy 1850 and 1858). Likewise, the contemporary editorial style was to cite only a person's last name within the text of an article. Thus, many references to "M. Darcy" (Monsieur Darcy) can be found in the French literature, where the writer implicitly assumed the reader knew Darcy's full name. Most of Darcy's associates may have never seen his name fully written out. Second and most importantly, his biographers Charié-Marsaines (1858) and Tarbé de St-Hardouin (1884) used "Henri". The second clearly took the spelling from the first, who may have copied it from a report written by the mayor of Dijon (Dumay 1845), or from Darcy's academic records from the Ecole

---

[1] Professor, Biosystems and Agricultural Engineering, Oklahoma State University, Stillwater, OK 74078; gbrown@okstate.edu

[2] Professor, VAW, ETH-Zentrum, CH-8092 Zurich, Switzerland

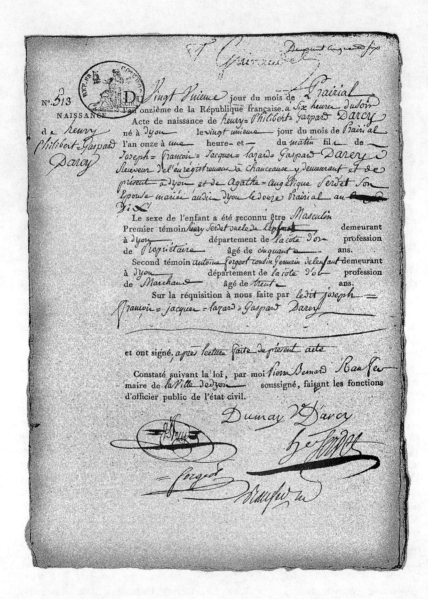

Figure 1. Henry Darcy's Birth Certificate.

Polytechnique (C. Billoux, Ecole Polytechnique Bibliothèque, Paris, "Responsable du service des archives", letter dated Oct. 24, 2000 to G. Brown.) However, the French spelling never appeared in Darcy's own works and his family has never used anything but "Henry" (P. Darcy 1957). Four later family descendants have carried the name and all have spelled it "Henry" (H. Darcy 2003).

To add to the confusion, some have spelled his last name "d'Arcy". While inappropriate to use now, he started his life with that noble spelling, as can be see in the family signature at the bottom of Figure 1. The signature is probably Darcy's father, and appears as "D'arcy" even though the official recorder uses "Darcy" elsewhere. After his father's death, Henry had a private tutor, who lent money to his mother. A Jacobin opposed to the aristocracy and all things aristocratic, he forced the boy to change the spelling to "Darcy" (P. Darcy 1957). (d'Arcy translates as "of Arcy" and implies nobility.) All in all, a particle was a small price to pay for an education.

Of final note, the signature "Dumay V" appears as a witness. This may well be Victor Dumay or his relative. Dumay would later serve as mayor and nominate Darcy to the Dijon Academy of Science, Art and Literature (Dumay 1845).

References

Charié-Marsaines, P. G. (1858). "Notice nécrologique sur M. Darcy, Inspecteur Général des Ponts et Chaussées". *Annales des Ponts et Chaussées*, Series 3; 15, 90-109.

Darcy, H. (1834). *Rapport à M. le maire et au conseil municipal, de Dijon, sur les moyens de fournir l'eau nécessaire à cette ville.* Douillier, Dijon.

Darcy, H. (1850). "Rapport à M. le Ministre des Travaux Publics, sur le pavage et le macadamisage des chaussées de Londres et de Paris". *Annales des Ponts et Chaussées*. Series 2, 10:1-264.

Darcy, H. (1856). *Les fontaines publiques de la ville de Dijon.* Dalmont, Paris.

Darcy, H. (1857). *Recherches expérimentales relatives au mouvement de l'eau dans les tuyaux*, Mallet-Bachelier, Paris.

Darcy, H. (1858). "Note relative à quelques modifications à introduire dans le tube de Pitot". *Annales des Ponts et Chaussées*, Series 3; 15, 351-359.

Darcy, H. (2003). "Henry Darcy: Inspecteur général des ponts et chaussées". *This volume*.

Dumay, V. (1845). *Notice historique sur l'établissement des fontaines publique de Dijon.* Frantin, Dijon.

Philip, J. R. (1995). "Desperately seeking Darcy in Dijon". Soil Sci. Soc. Am. J. 59, 319-324.

Tarbé de St-Hardouin, F. P. H. (1884). "Darcy". *Notices biographiques sur les ingénieurs des Ponts et Chaussées.* Baudry & Cie., Paris, 224-226.

Henry Darcy: Inspecteur général des ponts et chaussées

Henry Darcy[1]

Abstract

Henry P. G. Darcy is known for his achievements in engineering and contributions to his birthplace of Dijon, France. His descendant, Henry Darcy, the fifth person to carry the name, provides a family perceptive of his illustrious relative. In addition to biographical material, each of the other Henry's is described. Finally, excerpts from several letters from Darcy to Henry Bazin are used to show Darcy's modesty, humor and finesse. The complete paper is presented in both French and an English translation.

Figure 1. Henry Darcy at Ecole Polytechnique, circa 1821. (P. Darcy, 1957)

---

[1] Ingénieur civil des mines, Paris, France.

Introduction

L'importance de l'oeuvre scientifique et technique laissée par l'ingénieur Henry Philibert Gaspard Darcy justifie l'intérêt, voire l'engouement, que sa personnalité suscite aujourd'hui dans la communauté scientifique mondiale. Cet engouement se porte sur son oeuvre, bien sûr, mais également sur sa personnalité, sa vie et cette extraordinaire énergie qui lui permis en si peu de temps de faire autant de choses. Le « Darcy Memorial Symposium on the History of Hydraulics » est un signe de cet intérêt.

Les rédacteurs de ce volume ont souhaité qu'un témoignage familial soit apporté à ce séminaire. Notre famille a été très honorée de cette demande et c'est bien volontiers que j'apporterai ma contribution qui sera modeste compte tenu de l'abondance des informations disponibles aujourd'hui grâce au réseau internet dont les sites consacrés à mon illustre parent sont nombreux et actifs. Je vous proposerai donc un rapide rappel biographique de ce que fut la vie de l'ingénieur Darcy et je tenterai d'analyser ses principaux traits de caractères d'après ce qu'en rapporte la tradition familiale.

Héritage

Henry Philibert Gaspard Darcy est né à Dijon, capitale historique de la Bourgogne, le 10 juin 1803. Le nom de Darcy est ancien en Bourgogne puisqu'une lignée continue d'ecclésiastiques de haut rang le porte au XIII et XIV siècle en particulier dans la région d'Autun. Cette famille semble avoir été ruinée et dispersée lors des guerres de religion du XVI siècle au

Translation by Glenn O. Brown[1]
Introduction

The importance of the scientific and technical work left by the engineer Henry Darcy justifies the interest, even the passion, which his celebrity creates today in the world scientific community. This passion is based on his work of course, but also on his personality, life and extraordinary energy, which permitted great accomplishments in a short time. The "Darcy Memorial Symposium on the History of Hydraulics" is a sign of this interest.

The editors of this volume wished that our family's perspective be presented to the seminar, and we are greatly honored by this request. Thus, it is well that I contribute my share today, which will be modest taking into account the abundance of information available on the Internet, where many sites are devoted to my famous ancestor. Thus, I offer to you a small biographical note on the life of engineer Darcy. In it, I will examine his principal qualities according to our family tradition.

Heritage

Henry Philibert Gaspard Darcy was born in Dijon, historic capital of Burgundy, on June 10, 1803. Darcy is an old name in Burgundy. In particular, a continuous lineage of high-ranking clergymen carried it in the XIII and XIV century in the Autun region. The family appeared to be ruined and dispersed during the Wars of Religion in the XVI century. At that time, the family made the choice to reform as attested by old portraits of Luther and Calvin still appearing in the

---

[1] Professor, Biosystems and Agricultural Engineering, Oklahoma State University, Stillwater, OK 74078; gbrown@okstate.edu

cours desquelles elle fit le choix de la réforme comme en attestent d'anciens portraits de Luther et Calvin figurant encore dans le patrimoine familial au début du XIX siècle. L'ancêtre avéré et direct d'Henry Darcy, est le capitaine Pierre d'Arcy, son quart aïeul, né en 1618 et mort à Epinac en 1686. La famille s'était installée à Dijon au début du XVIII siècle et c'est là que naît en 1776 Jacques Lazare Gaspard d'Arcy, père du futur ingénieur. En 1794, c'est, en France, l'époque de la tourmente révolutionnaire. Lazare Gaspard s'enrôle dans le troisième bataillon des volontaires de la Cote d'Or; il participe aux combats historiques de Jemmapes et de Fleurus. C'est un « soldat de l'an II ». Comme l'écrivait un siècle plus tard son petit-fils,il appartenait à cette génération dont le rêve fut « La Révolution, la Grande Nation, la Grande Armée, la Liberté, la Dignité, la Félicité du genre humain par la France.» Le 28 Mai 1802, Lazare Gaspard d'Arcy épouse Agathe Angélique Serdet ; ils auront deux enfants : Henry, l'ingénieur, et Hugues Iéna, le préfet, mon quart aïeul.

Agathe Serdet était d'un milieu très modeste ; S'exprimant à propos de cette femme admirable qu'il avait bien connue puisqu'elle mourût en 1870 à 92 ans, mon trisaïeul, Henry II, son petit-fils, parvenu au soir d'une vie qui lui avait permis de côtoyer au Conseil d'Etat, dans la Préfectorale, et au sein du Patronat français tout ce que la France de l'époque comptait d'élites, n'en tirait aucune vanité et écrivait :

« A-t-il jamais passé du sang noble ou quart de noble par nos veines, je n'en sais rien ; mais pour sûr, souvenez vous-en mes enfants, il y a du sang de peuple, de ce bon peuple de France, de ce tout petit monde, point artiste ni compliqué, mais attaché en simplicité aux devoirs

family inheritance at the beginning of the XIX century. The direct ancestor of Henry Darcy was Captain Pierre d'Arcy, his grandfather fourth removed. He was born in 1618 and died in Epinac in 1686. The family settled in Dijon by the beginning of the XVIII century, and it is there that Jacques Lazare Gaspard d'Arcy, father of the future engineer, was born in 1776. In 1794 in France, revolution was stirring, and Lazare Gaspard enlisted in the Third Volunteer Battalion of the Cote d'Or. He took part in the historical battles at Jemmapes and Fleurus, and was a "soldier of the second year" [of the revolution]. One century later, his grandson wrote that Lazare Gaspard's generation's dream was "the Revolution, the Grand Nation, the Grand Army, Freedom, Dignity, and the Bliss of mankind through France." On May 28th, 1802, Lazare Gaspard d'Arcy married Agathe Angélique Serdet. They had two children, Henry the engineer, and Hugues Iéna the prefect, my grandfather, fourth removed.

Agathe Serdet had a very modest middle class background. He wrote about this admirable woman that he had known even though she was born in the previous century, since she died in 1870 at age 92. My great-great-grandfather, the second Henry, her grandson reached the end of a life where he dealt with the Council of State in the Prefecture, and within the French Council of Employers and gained everything that France of the time considered elite, but despite all this he was not vain.

"If ever passed noble blood or quarter of noble in our veins, I know nothing of it, but remember for certain my children, there is the blood of the good people of France. Not complicated artists, but simple people of a very small world devoted to daily duties, yet none the less

quotidiens ; non sans une pointe d'idéal cependant qui, aux heures bénies, parfume les destinées nationales, race de labour, mais travaillant comme chante l'alouette, gardant sous le hâle du jour toutes les sensibilités du coeur, secourable autant qu'économe, et en toutes choses, au naturel et sans pompe, amassant sou à sou le trésor matériel et moral du pays, et comblant sans jamais désespérer, les trous faits par les coquins, les dilettantes ou les sots. »

C'est à cette femme du « tout petit monde » qu'Henry et Hugues Iéna devront leur éducation et leur remarquable réussite. « Cette mère fut une éducatrice admirable. Elle était, elle est restée jusqu'à sa dernière heure débordante de tendresse vigilante, efficiente et désintéressée ; mais elle avait en même temps la volonté du but, le culte de l'effort et de la règle. Ce but qui est que, débiteur de tout ce qu'on vaut, on doit rendre tout ce qu'on peut. » Ces valeurs essentielles du travail, de l'effort et de la recherche du bien commun sont exactement celles que traduiront la vie et l'oeuvre de son fils Henry dont tous ses biographes ont souligné l'absence totale de recherche d'intérêt personnel, lui qui refusa toute rétribution particulière pour les services qu'il avait rendus, n'acceptant de la ville de Dijon que l'eau courante gratuite sa vie durant et une concession au cimetière municipal.

Lorsque Henry naît, en 1803, la France est en pleine épopée Napoléonienne et nul doute que sa famille suive avec passion l'actualité de l'époque : mon quart-aieul, né en 1807, se voit attribué le prénom de « Hugues Iéna » pour fêter la victoire que l'empereur a remporté sur les prussiens l'année précédente. Lazare Gaspard, étant fonctionnaire, on ne peut exclure totalement qu'un peu d'opportunisme

with ideals. Thus, at the proper time, they sense the national destiny and work hard but happily, keeping unto themselves all the sensitivities of the heart. In all things they are like a treasurer. With good nature and no conceit, they pile up penny upon penny, the material and moral treasure of the country. They fill without despairing, the holes made by the rascals, the dilettantes and the foolish."

It is to this woman of the "very small world" that Henry and Hugues Iéna will owe their education and their remarkable success. "This mother was an admirable teacher. She was and remained until her last hour, vigilant, efficient and unselfish in her tenderness. However, at the same time she had the will to reach a goal, with drive and determination. She was devoted to the belief that one owes all that she is worth, and one must return all one can." These essential work ethics, duty and the pursuit of the common good, are exactly those that her son Henry would translate into his own life and work. His biographers underlined the total absence of self-interested pursuits. He refused all commissions for the services that he rendered, and accepted from Dijon only free water service during his life and a concession in the municipal cemetery.

When Henry was born in 1803, France was in the Napoleonic era, and there is little doubt that his family passionately followed the popular movement of the time. My direct ancestor, born in 1807, saw himself given the first name "Hugues Iéna" to celebrate the emperor's victory over the Prussians the previous year. One cannot totally rule out that Lazare Gaspard, being a civil servant slipped a little political opportunism into the choice of the name. Under these conditions, one

politique se soit glissé dans le choix de ce prénom et on peut s'étonner dans ces conditions que Henry, né en 1803, ne se soit pas appelé Henry Marengo en l'honneur de la campagne d'Italie au cours de laquelle son oncle Jean d'Arcy avait été tué.

En 1817, Lazare Gaspard d'Arcy meurt en laissant une veuve et deux jeunes garçons dans une situation financière très précaire. Henry n'a que quatorze ans mais il se comporte alors comme un adulte responsable. Le soir de l'enterrement de son père, il prend à part son cadet Hugues Iéna, âgé de 10 ans, et lui dit « C'est notre père qui nous faisait vivre. Je ne t'abandonnerai jamais. Je serai le père. Aide moi en travaillant comme je travaillerai pour que, le plus tôt possible, nous gagnions notre pain et celui de notre mère avec honneur. (Darcy, 1957) »

Dès lors une exceptionnelle affection liera toute leur vie Agathe et ses deux fils. Dans une lettre écrite en 1823, Henry écrit à sa mère : « ne fondons notre bonheur que sur notre triple amitié » Ils resteront unis tous les trois toute leur vie et formeront ensemble, comme l'écrit Henry, une véritable « trinité ».

Ces deux enfants tinrent parole puisque le premier devint l'ingénieur que nous connaissons, et le second, mon quartaieul, devint un préfet important (Nîmes, Metz, Lyon) puis sous-secrétaire d'Etat à l'intérieur sous la seconde république (Forstenzer, 1981), avant de se reconvertir dans l'industrie en fondant la compagnie des forges de Châtillon et Commentry.

Agathe d'Arcy, n'ayant pas les moyens de financer les études de ses fils demanda et obtint des bourses de la ville de Dijon et un prêt de son beau-frère, par ailleurs tuteur des deux enfants : en

can wonder why Henry, who was born in 1803, was not called Henry Marengo in honor of the region in Italy where his Uncle Jean d'Arcy had been killed.

In 1817, Lazare Gaspard d'Arcy died and left his widow and two young boys in a very precarious financial situation. Henry was only fourteen years old, but he behaved like a responsible adult. The evening of his father's burial, he took his 10 year old sibling, Hugues Iéna, aside and said, "Our father supported us. I will never give up on you. I will be the father. Help me by working as I will, so that as soon as possible, we can earn our bread and that of our mother with honor (Darcy, 1957)".

Consequently, an exceptional affection bound Agathe and her two sons all their lives. In an 1823 letter, Henry writes to his mother, "lets found our happiness only on our triple friendship." All three remained united throughout their lives, and formed, as Henry wrote a true "trinity".

These two children kept their word, since the first became the engineer whom we know. The second, my ancestor, became an important Prefect (Nimes, Metz and Lyon) and then Under-Secretary of the Interior in the Second Republic (Forstenzer, 1981). He then changed occupations and founded the Forging Company of Châtillon and Commentry.

Agathe d'Arcy, not having the means to finance her sons' studies, asked for and obtained a scholarship from the city of Dijon and a loan from her brother-in-law, who was also a tutor for the two children. This republican brute advised the children to give up the particle and to transform d'Arcy into Darcy, which they did. Since that time, our family has chosen to preserve this corrected orthography because it is in this form that

contre partie, celui-ci, républicain farouche, conseilla aux enfants d'abandonner la particule et de transformer d'Arcy en Darcy, ce qu'ils firent. Notre famille a depuis lors choisi de conserver cette orthographe corrigée puisque c'est sous cette forme que notre nom a été illustré.

De même le « y » du prénom Henry tel que le portait l'ingénieur s'est transmis de génération en génération et je suis moi-même le $5^{\text{ème}}$ Henry Darcy depuis l'ingénieur.

La Lignée de Henry Darcy

I: Henry P.G. (1803-1858) l'ingénieur.

II: (1840-1926) Mon trisaïeul était le fils de Hugues Iéna. Il fit, comme son père, une carrière remarquable dans l'administration d'abord (préfet d'Epinal, d'Arras puis de Nice) puis dans l'industrie lourde (Charbonnages et Métallurgie) où il fonda et présida pendant quarante ans le Comité Central des Houillères de France (Isambert, 1965).

III: (1895-1916) Frère de mon grand-père, est mort pour la France à 20 ans.

IV: (1930-1953) Cousin germain de mon père était officier dans l'aéronavale et est mort en service commandé à 23ans.

V: (1954 -) Moi-même, mais revenons à l'ingénieur.

La Vie de Henry Darcy

En 1821, âgé de 18ans, il entre à l'école polytechnique qui était, et qui reste la plus prestigieuse des écoles d'ingénieurs françaises (Figure 1). En 1823 il complète sa formation à l'école des ponts et chaussées. A cette époque, il est décrit

our name was distinguished.

In the same way, the Henry first name carried by the engineer has been transmitted from generation to generation. I am the fifth Henry Darcy. [The later Henry's are not in a direct line and did not use middle names. They are denoted here by II, III, etc.]

The Henry Darcy Lineage

I: Henry P.G. (1803-1858) the engineer.

II: (1840-1926) My great-great-grandfather was the son of Hugues Iéna. Like his father, he had a remarkable career in administration. He was initially Prefect in Epinal, Arras and Nice. He then moved into heavy industry, (coal and metallurgy) where he founded and chaired for 40 years the Central Committee of the Collieries of France (Isambert, 1965).

III: (1895-1916) Brother of my grandfather, died for France at age 20.

IV: (1930-1953) The first cousin of my father was a naval aviator who died in service at age 23.

V: (1954 -) Myself, but let us return to the engineer.

Henry Darcy's Life

At age 18 in 1821, he entered the Ecole Polytechnique, which was and remains the most prestigious school of engineering in France (Figure 1). In 1823 he continued his education at the Ecole des Ponts et Chaussées (School of Bridges and Roads). At that time, he was described as having. "fire in his piercing eyes, and a fascinating smile." [The French is much more passionate.]

comme ayant « des yeux roux qui lançaient des flammes, ou s'enveloppaient dans un nuage farouche, quand ils ne s'éclairaient pas du plus vif,du plus affectueux et du plus prenant des sourires ».

En 1826 il rentre dans la vie professionnelle et, après un stage dans le Jura, est rapidement nommé à Dijon. En 1828, il épouse une anglaise de Guernesey, Henriette Carey, dont les parents habitent à Dijon, le bel hôtel Vogüe . De ce mariage, semble-t-il parfois difficile, ne naîtra malheureusement aucun enfant.

Puis viennent les trente années de vie professionnelles passionnantes et harassantes à la fois au cours desquelles Henry Darcy va littéralement brûler sa santé. Cette carrière est jalonnée de réalisations techniques et de découvertes scientifiques majeures.

Septembre 1840 : Alimentation en eau de la ville de Dijon qui devient, avec cent quarante bornes-fontaines réparties dans toute la ville, la ville d'Europe la mieux alimentée en eau après Rome, et bien avant Paris.

Juillet 1845 : Adoption par la chambre des députés du projet de tracer Darcy pour la ligne de chemin de fer de Paris à Lyon via Dijon. Ce tracé comporte le percement de tunnel de Blaisy (4,6 Km), ouvrage considérable pour l'époque.

1850-1852 : Mission à Londres pour l'étude de l'emploi du Macadam pour le revêtement des chaussées en remplacement du pavage, et à Bruxelles pour y étudier le projet d'alimentation en eau de la ville.

1856 : Publication de « *Les Fontaines Publiques de la ville de Dijon* » avec

In 1826 he began his professional life, and after a training course in Jura, he was quickly posted to Dijon. He married Henriette Carey in 1828, who was from the English island of Guernsey. Her parents lived in Dijon, at the beautiful Vogüe Hotel. It appears that the marriage was difficult, and unfortunately no children where produced.

Then came the 30 years of professional service that was both enthralling and difficult, during which Henry Darcy literally ruined his health. The career was marked by technical achievements and major scientific discoveries.

September 1840: Water supply of the city of Dijon is completed. With 140 public fountains, distributed throughout the city, it was the best water supply in Europe after Rome, but well before Paris.

July 1845: Darcy's plan for the Paris to Lyon railroad alignment through Dijon was adopted by the House of Commons. His design included the 4.6 km tunnel at Blaisy, a remarkable work for the time.

1850-1852: Mission to London to study the use of Macadamizing on roads [a graded stone foundation and pavement system]. He also traveled to Brussels to consult on the water supply of that city.

1856: Publication of *Les Fontaines Publiques de la ville de Dijon* [*The Public Fountains of Dijon*]. That publication contains "Darcy's law" for the permeability of porous media.

1857: Publication of *Recherches expérimentales relatives au mouvement de l'eau dans les tuyaux* [*Experimental research relating to the movement of water in pipes*], which provided the

l'énoncé de la « *loi de Darcy* » relative à la perméabilité des milieux poreux.

1857 : Publication de « *Recherches expérimentales relatives au mouvement de l'eau dans les tuyaux* ». Equation de Darcy-Weisbach.

2 Janvier 1858 : Henry Darcy meurt à Paris d'une pneumonie. Sa dépouille est transférée à Dijon par la voie ferrée qu'il avait dessinée ; à la gare, la troupe rend les honneurs et il reçoit l'hommage de toute la ville. Le lendemain, le conseil municipal de Dijon prend à l'unanimité la décision de donner son nom à l'actuelle Place Darcy qui marque l'entrée à Dijon des eaux de la source du Rosoir que Darcy avait canalisées.

Lettres à Bazin

L'ingénieur Darcy a entretenu une correspondance importante avec son jeune condisciple au Corps des Ponts et Chaussées, Henry Bazin ; ces lettres ne sont pas datées mais elles ont été numérotées de 1 à 25 par Bazin et sont très instructives quand au caractère de Darcy : elles sont empreintes de modestie, d'humour et de finesse.

Lettre n1 : « Et sur ce, je vous sers cordialement la main : On m'a écrit que vous aviez bu récemment à ma santé ; je crains que cette obligation ne vous soit longtemps encore imposée : mais le vin de Bourgogne est bon et je vous plains un peu moins de ce dévouement que de celui dont vous aurez besoin pour la lecture de ces épreuves. »

Lettre n7 : « Ce n'est pas dans un temps où l'on se joue si bien des principes, que je tiendrai, vous le pesez bien, à ma petite loi : a = α + β/R mais elle est si

[experimental] basis for the Darcy-Weisbach Equation.

January 2, 1858: Henry Darcy died in Paris of pneumonia. His body was returned to Dijon by the railway that he had designed. An honor guard met the train at the station, and he received the homage of the entire city. The following day, the Dijon municipal council unanimously decided to give his name to the current Place Darcy, which marks the entry terminal to Dijon of the Rosoir spring water that Darcy had developed.

Letters to Bazin

Engineer Darcy maintained a significant correspondence with his younger colleague from the Ponts et Chaussées, Henry Bazin. These letters are not dated but were numbered from 1 to 25 by Bazin. They are very instructive on Darcy's character, and are full of modesty, humor and finesse.

Letter 1: "And on this, I extend my hand cordially to you. It was written to me that you had toasted recently to my health. I fear that this obligation will be imposed on you for a long time, but the wine of Burgundy is good and I feel a little less sorry for this devotion than the one you will need to read these proofs."

Letter 7: "It is not in this time when people make light of the many principles, that I will hold on to it. But, weigh well my small law: $a = \alpha + \beta/R$ [Darcy's relation for friction coefficient in cast-iron pipe.], it is so simple and so convenient that I would like to see you make a final good effort in its favor "

Letter 8: "Finally you found my $\varepsilon = 0.00175$, which is about what I found

simple et si commode que je voudrais bien vous voir tenter un dernier effort en sa faveur. »

Lettre n8: « Enfin vous retrouvez mon $\varepsilon = 0.00175$; c'est à peu près ce que j'ai rencontré dans les tuyaux, peut-être trouverons-nous quelques procédés pour arriver à la valeur exacte ! Mais j'avoue en ce qui concerne ε que la foi est difficile, l'espérance est difficile, nous réclamerons donc de nos lecteurs, s'ils sont chrétiens, la troisième vertu théologiale. »

Lettre n15: « Je vous enverrai quand il les aura recopiées mes élucubrations sur la formule,
$$\varepsilon R^2(dv/d2) = ri/2$$
La raison de la chose m'échappe toujours, j'ai donc pris le parti d'affirmer qu'elle était vrai sur le blason de mon père. J'en ferai une affaire personnelle, et je trouverai bien quelques Durandal pour soutenir ma conviction : dans le temps où ma pauvre santé florissait, j'avais fait un pèlerinage à Rocamadour pour contempler cette glorieuse épée du neveu de Charlemagne. J'ai consacré, sans les regretter, 8 jours de ma tournée à la mémoire du gigantesque Roland : je préférerai cela à la lecture des livres de comptabilité et des livres de cantonniers. »

Le dévouement sans limites de l'ingénieur Darcy au bien public s'est clairement traduit par le sacrifice de sa santé. Il s'est, au sens propre, tué au travail. En 1855, il ne demande à être mis à disponibilité pour des raisons de santé que pour mieux se consacrer à ses recherches scientifiques et à ses publications capitales de 1856 et 1857. Il devait être déjà bien tard pour lui puisque la mort le prend dès Janvier 1858. Là encore, une lettre de Bazin illustre ce

in pipes. Perhaps we will find some way to arrive at the exact value! However in this, I acknowledge that the faith is difficult and the hope is difficult. We will thus ask our readers, if they are Christian, the third theological virtue [charity]."

Letter 15: "I will send you when it is copied my considerable discussion on the formula,
$$\varepsilon R^2(dv/d2) = ri/2$$
The justification of the thing [the formula] always escapes me. I have thus decided to affirm on the coat of arms of my father that it is true. I will make a personal quest of it, and I will find some Durandal [a famous sword's name] to support my conviction. In the time when my poor health flowered, I made a pilgrimage to Rocamadour to contemplate this glorious sword of Charlemagne's nephew. I devoted, without regretting them, eight days of my tour to the memory of the great Roland. I would prefer that to the reading of books on accountancy and road repair."

The unlimited devotion of Darcy to the public well being clearly resulted in the sacrifice of his health. He killed himself with work. In 1855, due to heath reasons, he asked to be relieved of duties to better devote himself to his scientific research and his major publications of 1856 and 1857. It may have already been too late, since death took him in January 1858. A Bazin letter illustrates his physical courage and brute energy.

Letter 21: "Night before last, I had nervous pains [in his body] so cruel that I screamed for 8 or 10 hours like a man under torture. Yesterday, the fatigue of the previous night gave me a violent migraine. Today, I am calm, but weak. Thursday, I hope to come and see you."

courage physique et cette énergie farouche.

Lettre n21: « J'ai eu l'avant dernière nuit des douleurs nerveuses si cruelles que j'ai crié pendant 8 ou 10 heures comme un homme à la torture : hier la fatigue de la nuit précédente m'a donné une violente migraine. Je suis calme aujourd'hui, mais faible. Jeudi, j'espère aller vous trouver. »

Remarques de Conclusion

Le plus grand écrivain et poète français du $19^{\text{ème}}$ siècle, Victor Hugo, né en 1802, était d'un an l'aîné de Henry Darcy. Les deux hommes partageaient sans doute un certain nombre d'idées, ils avaient de surcroît une référence commune: l'île anglo-normande de Guernesey: Henry avait épousé Henriette Carey dont la famille était originaire de l'île, Victor Hugo y fut exilé de 1851 à 1870. L'île est très petite et on peut rêver, rêver qu'à l'occasion d'une visite à son beau-frère qui y résidait, l'ingénieur et le poète se soient rencontrés et qu'ensemble ils aient marché le long de la grève et évoqués les grands sujets qui leur tenaient à coeur.

« J'ai bien souvent et bien sombrement rêvé » écrit Darcy à Bazin « dans mes longs jours et mes longues nuits de souffrances à ces mystérieuses causes finales dont Dieu seul s'est réservé la clef, et malgré mon horreur du doute, j'en suis réduit à douter encore. »

Références Bibliographiques

Darcy, P. (1957). *Henry Darcy, inspecteur général des ponts et chaussées*. Imprimerie Darantière, Dijon.

Concluding Remarks

Victor Hugo, the great French writer and poet of the 19th century, was born in 1802 and was one year older than Henry Darcy. The two men undoubtedly shared a number of ideals and they had a common connection to the Anglo-Norman island of Guernsey. Henry had married Henriette Carey whose family originated from the island, while Victor Hugo was exiled there from 1851 to 1870. The island is very small and one can dream that at the time of a visit with his brother-in-law, the engineer and the poet met. Together they may have walked along the shore and evoked the great subjects of their hearts.

"I very often have gloomily dreamed" Darcy wrote to Bazin "in my long days and long nights of sufferings to the mysterious final questions, who God alone reserves the key, and despite my aversion to uncertainty, I have little choice but to doubt still".

Forstenzer, T. R. (1981). *French Provincial Police and the Fall of the Second Republic.* Princeton University Press, Princeton, New Jersey.

Guyot, J. (1944). "L'Académie de Dijon et les ponts et chaussées". *Mémoires de l'Académie des Sciences, Arts et Belles-Lettres de Dijon*, 109:119-136.

"In memoriam". (1914). Bibliothèque Municipale, Dijon. [The anonymous author was a Darcy family member.]

Isambert, Y. (1965), "Darcy". *Dictionnaire de biographie française*, 10, 166-167, d'Amat, R. and Limouzin-Lamothe, R. (eds). Letouzey, Paris.

Jobert, P. (1991). *Les patrons du second empire*, Picard-Cénomane, Paris.

Henry Darcy's Perfection of the Pitot Tube

Glenn O. Brown[1]

Abstract

Starting in 1856 Henry Darcy, with the assistance of Henry Bazin, published four works that show various forms of an improved Pitot tube design. Although Henri Pitot had invented the device in 1732, theoretical and design weaknesses had kept it little more than a scientific toy. Darcy's improved instruments provided accurate and easy measurements of point velocity for the first time, which allowed advances in open channel and pipe flow hydraulics. His final design for the instrument tip is reflected today in all of our modern instruments. A reproduction of Darcy's published 1858 design was completed and shown to work as reported. Darcy's contribution to the development of the device equaled or exceeded Pitot's initial work, thus making it appropriate to refer to the modern instrument as the "Pitot-Darcy tube".

Introduction

The Pitot tube, is a simple and inexpensive instrument for the measurement of fluid velocity often taken for granted today. While largely replaced by rotating vane meters and various electronic instruments in hydraulic applications, it is still commonly used in pneumatic measurements and particularly for aviation airspeed determination. Its strength is of course its simple, robust design that allows an accurate velocity determination by measuring the pressure differential across two ports. While most practitioners appreciate the refinement of the instrument that occurred in the 20th century, many may not know its origin. This paper, which is a revised and expanded version of Brown (2001), will discuss the first instrument created by Henri Pitot, and the improvements made by Henry Darcy. In addition, basic testing is performed on a recreation of Darcy's design. It will be shown that Darcy, with the support of Henry Bazin, perfected the design into the useful instrument we use today.

---

[1] Professor, Biosystems and Agricultural Engineering, Oklahoma State University, Stillwater, OK 74078; gbrown@okstate.edu

Pitot's Instrument

Henri de Pitot (1695-1771), a French engineer and an early member of the *Corps des Ponts et Chaussées* (Corps of Bridges and Roads), made contributions in several areas of engineering, math and science (Chevray, 1969). However, his 1732 paper "*Description d'une machine pour mesurer la vitesse des eaux courantes et le sillage des vaisseaux*" (Description of a machine to measure the speed of running waters and the wake of vessels) is the highlight of his remembrance. Pitot's instrument consisted of two glass tubes mounted vertically on a wooden frame that had a length scale attached. The static tube pointed straight down, while the Pitot tube had a 90-degree bend at the bottom to face the flow. To obtain a measurement, the instrument was lowered into the flow and the difference in the liquid level in the two tubes recorded (Figure 1). Pitot described his excitement with the invention by writing,

> "The idea of this machine, known as a Pitot, is so simple and so natural that as soon as it had come to me, I ran at once to the river to carry out the first test with a simple tube of glass, and the effect met my expectation perfectly. After this first test, I could not imagine that such a simple, and at the same time very useful thing, had escaped so many skilled people who wrote and worked on the movement of water."

Pitot's device was recognized as innovative at the time and was described in detail by in the widely read thesis by Belidor (1737). However, it had four deficiencies that limited its application. First, Pitot did not provide the proper theoretical analysis for the device (Rouse and Ince, 1957). He equated the reading to the velocity attained by a falling body and prepared calibration tables of questionable accuracy. In fairness, while wrong his analysis was consistent with the practice of the day. Second, the instrument was slow and awkward to use. Third, the combination of the sizeable frame and the large static

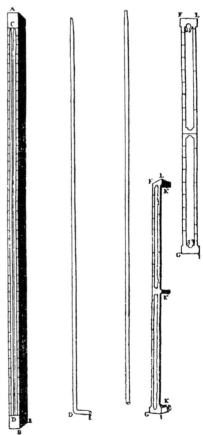

Figure 1. Henri Pitot's original design. (Rouse and Ince, 1957, reproduced by permission of IIHR.)

tube pointed downward distorted the flow and the precision of the static measurement. In fact, the water level in the static tube will in most cases be lower than the stream's water surface due to the fluid acceleration around its opening. Finally, all designs had trouble with oscillations in the water levels. There was no clear understanding of how the shape of the Pitot tube inlet affected performance. Some practitioners followed Pitot and used straight tubes, while others used tubes that had a funnel opening. An example of the latter is presented by Fanning (1877) (Figure 2). He writes, "The object of the expanded bulb and contraction below the bulb is to reduce oscillation of the water within the tube." The bulb obviously distorted flow around the device making its use problematic. Due to the instrument's weaknesses, it was little more than a scientific toy for the next 120 years (Hughes and Safford, 1926).

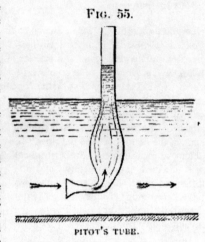

Figure 2. A funnel type Pitot tube (Figure 55, Fanning, 1877)

Darcy's Modifications

In 1858 another member of the French Corps, Henry Philibert Gaspard Darcy (1803-1858), published a paper that revolutionized Pitot's instrument and brought it into large-scale use (Darcy, 1858). The paper, "*Note relative à quelques modifications à introduire dans le tube de Pitot*" ("*Note relating to some modifications to be introduced to the Pitot tube*"), was published just after Darcy's death (Figure 3a).

Darcy noted the four weakness of Pitot's instrument and proceeded to correct each. First, while he diplomatically ignored Pitot's theoretical shortcomings, he provided a somewhat inelegant, but correct, analysis of the instrument's reading based on Torricelli's equation. His final result was,

$$V = \mu\sqrt{2gh} \qquad (1)$$

where V is the point velocity, g is the acceleration of gravity, h is the difference in the water levels and μ is a calibration coefficient dependent on the tip geometry.

The analysis allowed the pressure at the top of the manometer tubes to differ from atmospheric, which justified his modification to address the second problem. The tops of both tubes were connected through a valve to a short mouthpiece. The operator could place a small vacuum on both by opening the valve, sucking on the mouthpiece and then closing the valve. The vacuum would draw the water up the tubes to a convenient height for reading, but since the pressure on both was equal, the difference would remain the same.

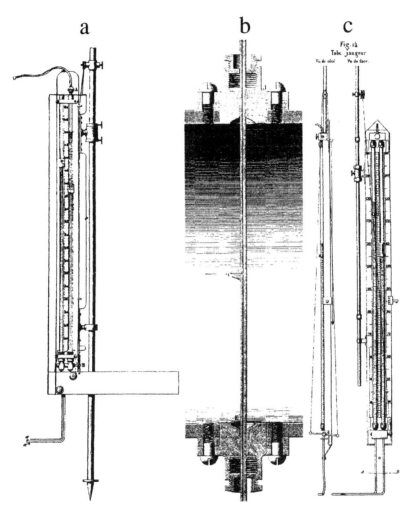

Figure 3. Darcy's Pitot tube designs by publication date;
a) 1858, b) 1857, c) 1856.

The third problem was addressed by replacing the bottom portion of the glass tubes with slender pipes that dropped clear of the support, and then bent forward. The Pitot tube of course faces the flow, while the static tube opened to the side. In addition, the tube support was tapered on both the leading and training edges. With these modifications in place Darcy stated, "Thus it is seen that the water is only very-imperceptibly disturbed at the point where the velocity is measured." Finally, his solution for the oscillation problem required two steps. He wrote,

"Then I made the oscillations in the tubes disappear almost entirely by making the openings only one and a half millimeters in diameter, while that of the tubes was one centimeter. As these oscillations, however reduced they were, could still confuse the operator; I placed a valve so that one can simultaneously close the lower openings of the tubes; these openings being closed, any communication with the current is stopped, and one can read the difference to deduce the velocity with ease and precision."

This paper was neither the first nor last of Darcy's publications that addressed the use of Pitot tubes. A comparison between these publications shows a steady advance in instrument design. Darcy's first use of the device (Darcy, 1857; Gisonni, 2003) was in the measurement of the pipe flow velocity distributions. (The actual work was done from 1850 to 1854.) Figure 3b shows the Pitot tube that he installed in pipe test sections. The tube could slide up and down and fit into a recess in the pipe wall, while the static line was installed flush on the pipe wall. This device displays a tapered tip and valves to stabilize the readings consistent with the later designs. It is quite possible that the valves were installed for operational reasons, and their use in measurement came later. Darcy (1856) shows a very similar design to the 1858, with the exception of the tip. In Figure 3c, the string and pulley apparatus that allows both lower valves to be closed simultaneously from above can be seen. The last publication, (Darcy and Bazin, 1865; Hager and Gisonni, 2003) again showed a design similar to the 1858 device with a third tip design. Henry-Emile Bazin (1829-1917) was Darcy's protégé and assisted him for several years, and Darcy acknowledged his work on the testing of the Pitot tube (Darcy, 1858). The 1865 publication presented the results of open channel experiments originally designed by Darcy, but Bazin undoubtedly wrote the entire document.

Figure 4 presents the three instrument tip variations. The 1856 drawing shows the Pitot tube lengthened considerably compared to the pipe-flow device (Figure 4a). More importantly, the static line was placed next to the Pitot line, and its opening formed by a sideways 90° bend. In 1858, the static pressure line opens with what appears to be a soldered fitting mounted on the bottom, but its details are neither clear in the text nor in the illustration (Figure 4b). Finally, the 1865 report presents a streamlined assembly very similar to modern designs (Figure 4c). However, the publication dates of the three tip designs are probably misleading, as to when they were actually in service. Darcy (1856) describes three calibration tests with different static port designs, and in a footnote, Darcy (1858) seems to describe the 4b design as a fabrication error, which he replaced quickly with 4a. Similarly, in the discussion of calibration coefficients, he states the 4a design has a calibration coefficient, $\mu = 0.84$, while another tip with the static port formed by simply piercing the tube wall had a coefficient of 1. Only a design similar to 4c could have had a coefficient with that

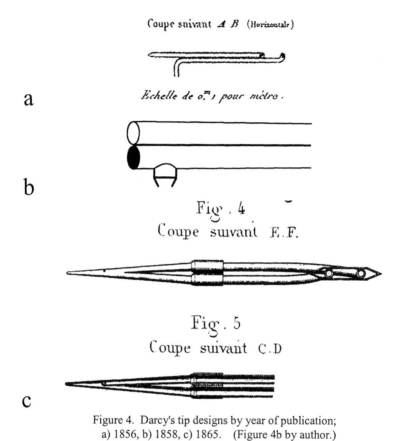

Figure 4. Darcy's tip designs by year of publication;
a) 1856, b) 1858, c) 1865. (Figure 4b by author.)

value. Since the 1858 paper was published after his death, it is possible that the drawing of the older instrument was used by the editors, instead of the most recent design.

When closing the topic, Darcy (1856) wrote the rather prophetic statement; "It is essential to notice that a seemingly unimportant modification in the form or the provision of the second [static] tube can have a great influence on the value of μ". This fact of course leads to a variety of subtle design modifications through the years (Peters, 1931), with the best generally credited to Ludwig Prandtl (1875-1953) (Merriam and Spaulding, 1935). His design called for a rounded tip Pitot tube to project 11 to 13 tube diameters ahead of the tubing bend, and the static port to be a continuous ring, three tube diameters behind the Pitot port. As can be seen, the tip shown in Figure 4c approaches that design.

Instrument Replica

In an effort to gain a better understanding of Darcy's design, a replicate of the 1858 instrument was constructed (Figure 5). If the 1858 paper did not provide design guidance for a given feature, the other reports were consulted. In the end, several details had to be inferred, but it is believed that the result is a fairly truthful model. An initial difficulty was in maintaining a period look with modern materials. In particular, the lower valves were troublesome to duplicate. No stock designs could be located that would allowed two valves to be operated by a single level. As a result, it was necessary to rework two old brass stopcocks into a single assembly. The final instrument was 191 cm tall, 47 cm deep across the rudder and 4 cm wide at the manometer board. The extensive use of brass fittings and a steel pipe (2.1 cm OD, 1.6 cm ID) produced a relatively hefty 6.0 kg apparatus.

Figure 6 shows the device in a flume with a water velocity of 0.3 m/s. As can be seen, the design was effective in minimizing flow disturbance. Rather humbling for the author was the difficulty encountered in getting the vacuum differential to work properly. *Any* leak in the system would result in erroneous readings. Thus, all fittings and connections had to be tight fitting, but resilient to operational shocks, which proved to be difficult to achieve. The glass manometer tubes broke more than once, and their rigid connection to the lower valve had to be replaced with flexible shrink tubing. Likewise, the lower brass stopcocks had to be replaced with very un-period looking modern valves before the instrument would work properly. Once

Figure 5. Replica of 1856 Pitot-Darcy tube.

Figure 6. Replica of 1858 design in testing.

everything was tight, it worked well and the calibration obtained was similar to that reported by Darcy. Overall, the experience impressed upon the author the obvious skill of the 19th-century instrument makers who made the original.

Concluding Remarks

Darcy's improvement of the Pitot tube was soon put to work by other researchers and contributed to the rapid advances in hydraulics of the late 19th and early 20th century. An example of data being collected with the 1865 design is presented in Figure 7 (Darcy and Bazin, 1865), while Figure 8 shows Bazin's plotting of the isovels obtained in a rectangular channel. Integration over the area allowed accurate estimates of volumetric flow and channel resistance. Similar detailed analysis of pipe flow isovels (Darcy, 1857) allowed improvement in close conduit friction equations.

Figure 7. Technicians using a Pitot-Darcy tube (Darcy and Bazin, 1865).

Finally, it should be noted that Darcy unselfishly gave the instrument to the world. Charie-Marsaines (1858) wrote,

> "Perfections brought by Darcy to this instrument are considerable and he would have been able to take a patent to profit from its exclusive manufacture during a certain number of years. However, as a believer in the customs of disinterestedness, he decided to abandon his invention to the public. The administration has already made a number of these instruments some of which are placed in the precision instrument deposit at *l'École des Ponts et Chaussées*, and others sent to state engineers to be used in operations that demand exact measurements of the water velocity."

Clearly, Darcy's contribution to the development of the device equaled or exceeded Pitot's initial work. Thus it is only right, as some authors do now, to refer to the modern instrument as the "Pitot-Darcy tube". With this nomenclature, the dynamic pressure is measured with the "Pitot" tube, while the hydrostatic pressure is obtained with the "Darcy" tube.

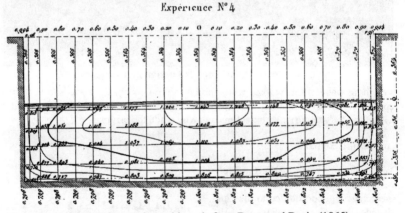

Figure 8. Example of open channel isovels from Darcy and Bazin (1865).

References

Belidor, B. F. de, (1737). *Architecture Hydraulique*, Volume 1, Jombert, Paris.

Brown, G. O. (2001). "Henry Darcy and the Pitot tube", in *International Engineering History and Heritage*, J. R. Rogers and A. J. Fredrich eds, ASCE, Reston, VA, pg. 360-366.

Charié-Marsaines, P. G. (1858). "Notice necrologique sur M. Darcy, Inspecteur Général des Ponts et Chaussées", *Annales des Ponts et Chaussées, Series 3; 15*, 90-109.

Chevray, R. (1969). "A man of hydraulics: Henri de Pitot (1895-1771)". *Journal of the Hydraulics Division*, ASCE 95(HY4), 1129-1138.

Darcy, H. (1856). *Les fontaines publiques de la ville de Dijon*. Dalmont, Paris.

Darcy, H. (1857). *Recherches experimentales relatives au mouvement de l'eau dans les tuyaux*, Mallet-Bachelier, Paris.

Darcy, H. (1858). "Note relative à quelques modifications à introduire dans le tube de Pitot". *Annales des Ponts et Chaussées*, Series 3; 15, 351-359.

Darcy, H. and Bazin, H. (1865). *Recherches Hydrauliques, enterprises par M. H. Darcy*, Imprimerie Nationale, Paris.

Fanning, J. T. (1877). *A practical treatise on water-supply engineering*, Van Nostrand, New York.

Gisonni, C. (2003). "Henry Darcy and the pipe flow formula", *this volume*.

Hager, W. H. and Gisonni, C. (2003). "Henry Bazin: Hydraulician", *this volume*.

Hughes, H. and Safford, A. (1926). *A Treatise on Hydraulics*, Macmillan Company, New York.

Merriam, K., and Spauliding, E. (1935). *Comparative tests of Pitot-static tubes*. N.A.C.A. Tech. Note 546, Washington, D.C.

Peters, H. (1931). "Druckmessung". *Hydro-und Aero-Dynamik 1: Strömungslehre und allgemeine Versuchstechnik*, L. Schiller, ed. Akademische Verlagsgesellschaft, Leipzig, pg. 487-510.

Pitot, H. (1732). "Description d'une machine pour mesurer la vitesse des eaux courantes et le sillage des vaisseaux", *Mémoires de L'Académie*, November.

Rouse, H. and Ince, S. (1957). *History of Hydraulics*, Iowa Institute of Hydraulic Research, The University of Iowa, Iowa City.

Henry Darcy and the Pipe Flow Formula

Corrado Gisonni[1]

Abstract

Despite a relatively short life, Henry Darcy (1803-1858) was able to leave an unforgettable legacy of engineering activity, as well as research contributions in hydraulics. His major findings are well known worldwide: The basic law for groundwater flow, the flow resistance formula for pipes and improvements to the Pitot tube. The present paper focuses on the pipe flow experiments, which were conducted by Darcy at Chaillot in Paris from August 3, 1849 to October 27, 1850. At that time Darcy was in charge as Chief Engineer of the municipal service of Paris. Details of his experimental research, along with a lucid presentation of the state of the art in pipe flow resistance formulas, may be found in his wonderful "Mémoire", submitted by Darcy to Académie des Sciences in June 1854 and published in 1858.

Introduction

On the occasion of the bicentennial of his birth on the 10th of June, 2003, it is appropriate to celebrate Henry-Philibert-Gaspard Darcy, renowned pioneer of hydraulic research (Brown et al. 2000). Darcy's name is associated with several different topics in hydraulic engineering and the relationships comprising part of his scientific legacy are used daily. These are notably;
- Darcy's law of groundwater flow, which states that the average cross-sectional velocity across a porous media is equal to the hydraulic conductivity times the hydraulic gradient;
- The Darcy-Weisbach equation, which states that the friction slope in pipe flow is equal to the velocity head times the friction coefficient divided by the pipe diameter; and
- The Darcy number (or Darcy-Rayleigh number) that applies to heat transfer in flow through porous media.

Two virtues facilitated Henry Darcy's work. He was a talented engineer and a gifted researcher. As a consequence, he was able to conceive two extraordinary experimental installations: One at Chaillot in Paris to investigate pipe flow (Darcy

---

[1] ASCE Member, Associate Professor, Dipartimento di Ingegneria Civile, Second University of Naples, via Roma 29 – 81301 Aversa (CE), Italy. Phone +39 081 5010220; corrado.gisonni@unina2.it

1858a), another at Canal de Bourgogne in Dijon for open channel experiments in cooperation with his assistant Henry-Emile Bazin (1829-1917) (Bazin 1865). The present paper focuses on the first work, whereas the second work is discussed in Hager and Gisonni (2003).

Au mouvement de l'eau dans les tuyaux

In 1858 l'Académie des Sciences de l'Institut Impérial de France issued the fifteenth volume of Mémoires, which included six works from various scientists. Mixed among different topics (biology, ophthalmology, and chemistry) it is possible to read (page 141) the following title (Figure 1).

Recherches expérimentales relatives au mouvement de l'eau dans les tuyaux
par M. H. Darcy
Inspecteur Général des Ponts et Chaussées

Unfortunately, Henry Darcy never had the chance to see that volume, which was published after his death, given that the review process and the printing procedure took about four years. It was probably the long delay by the Académie des Sciences that convinced Darcy to prepare a concise resume of his main results on the pipe flow research. It was included as Chapter II of Part III of Les fontaines publiques de la ville de Dijon (Darcy 1856).

The draft was submitted to the Académie des Sciences during the first half of 1854, according to its reviewers (Poncelet et al. 1854). The exceptional review committee was composed of Jean-Victor Poncelet (1788-1867), predecessor of Adhémar Jean Claude Barré de Saint-Venant (1797-1886) in the mechanics section of the Académie des Sciences; Charles Combes (1801-1872) long time president of *Académie des Sciences*, and Arthur Morin (1795-1880), a well known successor of Poncelet at Metz and later a general in the French Army.

The paper (Darcy 1858a) of 262 pages is subdivided into six chapters and four appendices, with the table of contents;
- Chapter I: state-of-the-art on pipe flow,
- Chapter II: description of the experimental set up,
- Chapter III: presentation of experimental results,
- Chapter IV: evaluation of empirical coefficients for head loss formula,
- Chapter V: velocity distribution in pipe flow,
- Chapter VI: flow contraction at the pipe inlet,
- Note 1: description of equipment for pressure measurements,
- Note 2: description of device for velocity measurement (Pitot tube),
- Note 3: similitude between pipe flow and open channel flow,
- Note 4: resume of experimental tests, and
- Tables for practical computation of discharge in new cast iron pipes (diameters ranging from 0.01 to 1.00 m) depending on different values of velocities (ranging from 0.10 to 3.00 m/s).

The following sections discuss each of these chapters.

RECHERCHES EXPÉRIMENTALES

RELATIVES

AU MOUVEMENT DE L'EAU

DANS LES TUYAUX,

PAR M. H. DARCY,

INSPECTEUR GÉNÉRAL DES PONTS ET CHAUSSÉES.

SOMMAIRE.

Ce Mémoire traitera du mouvement de l'eau dans les tuyaux. Il est divisé en six chapitres.

Le premier chapitre a pour objet d'indiquer les motifs qui m'ont déterminé à me livrer à ces recherches expérimentales.

Le second chapitre est consacré à la description des appareils employés dans ces recherches.

Le troisième présente le résultat des expériences.

Le quatrième, les procédés employés pour déterminer les coefficients des formules.

Le cinquième donne la description des expériences relatives à la recherche de la loi qui lie entre elles les vitesses des filets fluides.

Dans le sixième et dernier, je détermine le coefficient de contraction à l'entrée des conduites cylindriques; je fais précéder cette recherche d'un résumé des résultats obtenus dans le cours de ce Mémoire et de quelques considérations relatives aux variations respectives des deux termes de la résistance *dans l'expression générale* d'où l'on déduit la vitesse moyenne de l'eau dans une conduite cylindrique.

Figure 1. Title page (Darcy, 1858a).

Chapter I. The first chapter illustrates the current knowledge on pipe flow. Darcy first emphasized the scarcity of experimental data used as a basis for calibration of flow resistance formulas by his predecessors. Second, he showed the reasons why it was worthwhile to conduct such research. Prior to Darcy, few experiments on pipe flow were available. In fact, the original formula proposed by Gaspard-Marie Riche de Prony (1755-1839) was essentially based on fifty-one experiments. These consisted of seven tests using the conduits installed at Versailles conducted by Claude Antoine Couplet (1642-1722), twenty six tests in iron tinplate pipes made by Charles Bossut (1730-1814) and eighteen tests in iron tinplate pipes carried out by Pierre Dubuat (1734-1809). De Prony presented his flow resistance formula as,

$$\frac{1}{4}Dj = \alpha \ v \ + \beta \ v^2 \qquad (1)$$

where D is the pipe diameter (m), j is the unit head loss, v is the average velocity (m/s), and α and β are constant coefficients set to 0.0000173314 and 0.0003482590, respectively.

De Prony's formula was the key point during hydraulics lectures at the prestigious Ecole des Ponts et Chaussées at Paris, despite that its practical application was criticized by eminent engineers. Darcy referred to a letter of Jean-François d'Aubuisson (1769-1841), designer of the water supply system of Toulouse, stating that Eq. 1 was derived by experiments using small diameters. Consequently, it was not reliable for hydraulic design. D'Aubuisson mentioned various fountains in Paris providing only half of the discharge expected by municipal engineers.

Another questionable point was that de Prony never considered the influence of pipe roughness on the friction. Following his interest in solving practical problems, Darcy decided that it was mandatory to start experiments with two main objectives. To quantify the dependence of discharge on specific wall surface conditions of pipes and to determine the influence of pipe diameter on flow resistance.

In a footnote of Chapter I (p. 151) the author acknowledged Poncelet and Joseph-Baptiste Bélanger (1790-1874) for having attended several experimental runs and for providing structural suggestions.

Chapter II. A complete and exhaustive presentation of the experimental site at Chaillot in Paris is presented within the second chapter. Tests started on the 31st of August 1849 and terminated on the 27th of October 1851, while Henry Darcy was director of the Municipal Water Service of the French capital. Despite an ambitious goal, Darcy did not conceal his optimism: "Tout se réunissait pour faciliter mes recherches." (Everything combined in order to facilitate my research). Figure 2 gives an idea about how Chaillot must have appeared in Darcy's times. The view is looking downstream along the Seine, the Palais de Chaillot is clearly observable on the right river bank. (The future site of the Eiffel tower is out of the field of vision, but is located just on the opposite river bank.)

The location of Chaillot offered the best combination of natural and technical resources. The hill slope allowed considerable head on experimental pipes, a large amount of water was readily available from the Seine thanks to pumps lifting water up to the supply reservoir, large existing equipment were available to measure discharge, and capable and devoted personnel from the municipal staff provided

Figure 2. Palais de Chaillot in Darcy's Age.

support. Beautiful plates, separated from the text, illustrate the organization of that unique "open air laboratory".

In addition, the experimental site was close to the Chaillot industrial area where mechanical industries were in operation. Darcy profited of such a good opportunity and often used local facilities for maintenance and repair of pumps and other equipment. Unfortunately, early in the morning on the 16th of December 1865, a large fire destroyed more than two thirds of the industrial area and probably also involved the experimental facilities. Nowadays there are no remnants of Darcy's installation, the district of Chaillot is now residential and streets have been renamed (e.g. the experimental pipe was installed along rue Basse Saint Pierre de Chaillot, which is now rue de la Manutention).

With regard to technical details, Figure 3 shows Darcy's list of tests. Diameters ranged from 0.0122 to 0.50 m, while pipe materials included glass, iron, lead, bitumen coated iron, tinplate and cast iron. Wall conditions varied from new to used with deposits, and average velocities ranged between 0.03 and 6.01 m/s. Discharges were measured by collecting water in tanks of defined volume. Lengths of pipe were always more than 100 m, except for the lead and glass pipes, being 50 and 44.8 m, respectively.

Pressures were measured by means of five manometers (Figure 4) installed at different locations. The first manometer was installed at the feeding reservoir, the second at the pipe inlet section, the third about 4.70 m downstream and the fourth and fifth were spaced 50 and 100 m from the third. Darcy used only pressure data read at the last three manometers to quantify the head loss in order to avoid inlet effects. His data also included joint effects, exactly numbered for each type of conduit (Figure 3). In total, 198 experiments were performed, or almost four times the number in the database used by de Prony, in developing his formula.

Chapter III. The experimental results are illustrated in this chapter and include ten tables on the original data sheets. The precision of the data tables is astonishing. They even state the recording date, the starting time and ending time for each run. The tables also list comparison with predictions from de Prony's formula, which are

152   DU MOUVEMENT DE L'EAU

TABLEAU DES TUYAUX SOUMIS AUX EXPÉRIENCES.

NUMÉROS D'ORDRE.	DIAMÈTRES.		NATURE DES TUYAUX.	ÉTAT des SURFACES.	LONGUEUR de chaque tuyau.	NOMBRE DE JOINTS par 100 mètres.	MODE DE RACCORDEMENT.	VITESSES MOYENNES OBTENUES.	
								Minimum.	Maximum.
	mèt.				mèt.			mèt.	mèt.
1	0,0122	⎫	For étiré	neuf	2,15	47	A vis	0,0344	1,195
2	0,0266	3		idem	2,47	40	Idem	0,0578	2,1840
3	0,0395	⎭		idem	3,85	26	Idem	0,0626	2,3071
4	0,014	⎫	Plomb	neuf	11,75	8	Nœuds en soudure	0,04	1.29
5	0,027	3		idem	6,60	16	Idem	0,065	1,679
6	0,041	⎭		idem	5,20	20	Idem	0,12	2,305
7	0,0268	⎫	Tôle et bitume	neuf	2,79	36	A vis	0,03	2,507
8	0,0826	4		idem	2,90	34	Idem	0,10	3,897
9	0,196			idem	2,90	34	Idem	0,18	6,01
10	0,285	⎭		idem	2,90	34	Idem	0,395	3,207
11	0,04968	1	Verre	neuf	1,16	86	Joints à brides	0,153	2,108
12	0,0359			avec dépôts	1,30	77	Idem	0,051	0,033
13	0,0364			nettoyé	1,30	77	Idem	0,113	1,126
14	0,0795			avec dépôts	2,50	40	Emboîtements	0,123	1,142
15	0,0801			nettoyé	2,50	40	Idem	0,193	1,526
16	0,0819			neuf	2,50	40	Idem	0,088	3,265
17	0,0137	11	Fonte	idem	2,50	40	Idem	0,140	4,693
18	0,188			idem	2,50	40	Idem	0,205	4,928
19	0,2432			avec dépôts	2,50	40	Joints à brides	0,307	3,833
20	0,2447			nettoyé	2,50	40	Idem	0,278	4,497
21	0,207			idem	2,50	40	Emboîtements	0,244	3,16
22	0,50	⎭		neuf	2,50	40	Idem	0,4207	1,1278

Figure 3. List of Experiments (Darcy, 1858a).

not encouraging. Errors in velocity ranged up to 33 %. Lead pipes with 14, 27 and 41 mm of diameter provided the best agreement.

The first substantial evidence for the need for a new formulation emerges quickly from the experiments. Pipe material and conditions of the pipe wall significantly influence the head loss in pressurised flow! Other striking indications were,

- the pipe diameter had definite influence on head loss, given that Eq. 1 overestimates head losses for small diameters and underestimates for larger diameters, and
- For velocities lower than 0.10 m/s resistance is linearly proportional to the average velocity, because the quadratic term of Eq. 1 becomes significant.

Chapter IV. This is the core of Darcy's work on pipe flow. After a careful observation of phenomena, Darcy proposed that the experimental data could be properly fitted by,

$$R\ i = a\ v\ + b\ v^2 \tag{2}$$

or, more suitable for practical computations as

$$R\ i = b_1 v^2 \tag{3}$$

where R is the pipe radius, i the unit head loss and a, b and b_1 the coefficients describing pipe material and conduit age, to be determined through the least squares

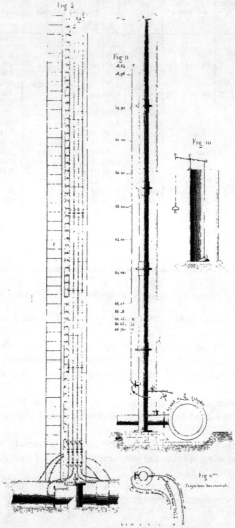

Figure 4. Typical manometer (Darcy, 1858a).

method. Equation 3 is recommended for practical purposes also by Poncelet et al. (1854), and the coefficient b_1 was expressed as

$$b_1 = \alpha + \frac{\beta}{R} \qquad (4)$$

with α equal to 0.000507 and β equal to 0.00000647, for cast iron pipes and diameters up to 0.50 m. Finally it was recommended to double the value of b_1 to account for the presence of deposits or the aging of the conduit.

Figure 5 shows a cast iron pipe installed in the water reservoir at *Place Darcy* in Dijon, a key structure of the water supply system designed by Henry Darcy for his native town. The water supply system for the city of Dijon is definitely his masterpiece in terms of hydraulic infrastructure (Hager and Gisonni 2002, Gisonni 2002). Darcy was able to address the destiny of that city, capital of Duchy of Burgundy, which was an ill-famed place because of the poor quality of its water, at the beginning of the 19th century (Darcy 1834, Darcy 1856). The pipe of Figure 5 was certainly designed using Eq. 3. Once again a wonderful proof of practical application of scientific speculation.

In order to make Eq. 4 readily applicable to practitioners, Darcy prepared tables, based on Eq. 3, with diameter and velocity varying respectively from 0.01 to 1.00 m, and from 0.10 to 3.00 m/s. Equation 3 may be rewritten as

$$i = \left(0.00164 + \frac{0.000042}{D}\right)\frac{Q^2}{D^5} \qquad (5)$$

Equation 5 is definitely a milestone in the history of hydraulics, because it corresponds to an indispensable tool for engineers dealing with water supply systems.

Chapter IV ends with another impressive statement. Flow resistance in smooth pipe with velocities smaller than 0.10 m/s can be considered as linearly dependent on average velocity, because the quadratic term in Eq. 2 becomes insignificant. In principle, such flow follows the groundwater arrangement, as expressed by the Darcy law for groundwater flow. Although Darcy never formulated the difference between laminar and turbulent pipe flows, he was definitely aware of its practical implications.

J. T. Fanning (1837-1911) was the first to effectively combine Darcy's experimental results with Weisbach's equation form. Like Darcy, he provided tables where different values of the friction factor f were indicated as a function of pipe material, diameter and velocity (Brown, 2002).

Chapter V. The velocity distribution in pipe flow is tackled in Chapter V. Here Darcy intended to answer the question; What is the law governing the distribution of velocities in terms of the radial coordinate, given that maximum and minimum values are respectively located at the axis and at the pipe wall?

Local velocities in pipes were measured by means of a special Pitot tube installed inside the experimental conduits (Figure 6). Henry Darcy worked for a long time on that device, looking for continuous improvement. At that time, technicians knew of the original instrument conceived in 1732 by Henri de Pitot (1695-1771). However, it was considered nothing but a theoretical speculation, without interest for

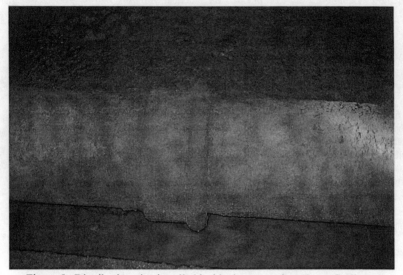

Figure 5. Distribution pipe installed inside the reservoir at *Porte Guillaume*.

practical applications (Combes 1857, Darcy 1856, 1858b and 1858c, Brown 2001). Darcy wrote various papers on his velocity meter, and even a large portion of Appendix C of *Les fontaines publiques de la ville de Dijon* (Darcy 1856) refers to the *Principe et description du nouveau tube jaugeur*. That article is essentially a resume of Darcy's improvements to the Pitot tube. Probably he was afraid that his work would not be published before his death, because of the usual protracted time in the review process and other inconveniencies. Darcy's contribution to the development of a better device for velocity measurement is so important that it is only right to refer here to the *Pitot-Darcy tube*. From interpolation of experimental data, Darcy came to the following relation for velocity distribution

$$V - v = K \frac{r^{\frac{3}{2}}}{R} \sqrt{i} \tag{6}$$

where V is maximum velocity, v is the local velocity at distance r from the pipe centre, and K as the experimental coefficient equal to 11.30. Figure 7 shows a comparison between computed values from Eq. 6 (continuous line) and measured values for various discharges.

In accordance with Eq. 6, Darcy computed the velocity w at the wall ($r = R$) and the average velocity u, respectively, as

$$w = V - K \sqrt{R\ i} \tag{7}$$

$$u = \frac{3\ V + 4\ w}{7} \tag{8}$$

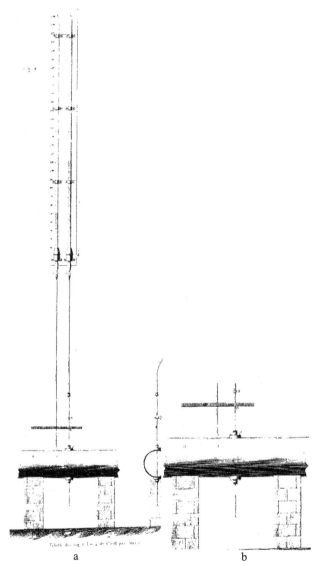

Figure 6. Velocity measurement equipment; (a) general view, (b) detail

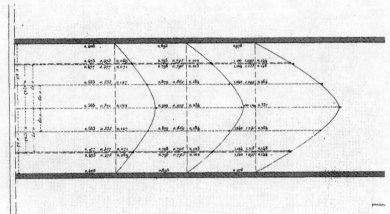

Figure 7. Velocity distribution plot.

Note, Darcy identified the average velocity with the symbol u when referring to velocity distribution, whereas he used the symbol v in the flow resistance formula. Using these equations he derived that u is located at $r = 0.689\,R$, as confirmed by experimental results. With the velocity distribution Darcy was also able to compute the energy correction coefficient (or Coriolis coefficient) and the momentum correction coefficient (or Boussinesq coefficient) to demonstrate that those can be assumed equal to unity in practice.

The weakest point of Eq. 6, and thus also in Eq. 7, is obvious. The wall velocity was not zero. At that time, non-zero values were commonly assumed, as evidenced that the eminent reviewers supported the approach. The hydraulic community had to wait for more than seventy years, until the results of Ludwig Prandtl (1875-1953) and Theodor von Karman (1881-1963) became available.

Chapter VI. Reviewers heavily criticized Darcy's results on the contraction coefficient at a pipe inlet. In fact the committee clearly stressed that this topic had already been tackled by Poncelet in 1841, and wrote that "only by compensation of errors he (Darcy) obtained the same average value of the contraction coefficient at pipe inlets, usually assumed equal to 0.825" (Poncelet et al. 1854).

Of course, Darcy was not happy with that comment, but his personality was strong enough to take the blow. In the last pages of Chapter VI, the engineer and scientist from Dijon stated precisely that he should have shortened the *Mémoire* but intended to answer punctually to the reviewers' comments. Therefore, after detailed explanations on his measurements of pipe inlet effects, Darcy stated that this phenomenon had an irrelevant influence in any case on the practical computation of head loss in pipe flow.

The Appendix (Notes 1 to 4) consists of ten pages in total (pp.360-369). They contain no special information except for Note 3, where Darcy mentioned his experiments in rectangular channels, by then started with the support of the young and capable engineers André Baumgarten (1808-1859) and Charles Ritter (1825-

1902), and concluded with the cooperation of Bazin (then in charge at the *Canal de Bourgogne*, in Dijon). Poncelet et al. (1854) highly encouraged publication of that research in order to complete Darcy's scientific mission. As is well known, the final results on the open channel investigations were published seven years after his death (Bazin 1865).

During his career Henry Darcy was often supported by excellent disciples, the most famous of all being undoubtedly Bazin. Darcy had an excellent ability as team leader; he constantly managed his staff as a friendly association based on mutual respect, as is testified by private correspondence between Bazin and himself (Gisonni and Hager 2003). Some time ago a notable Italian scientist said: "There are two types of scientific leaders: One that becomes important by trampling on his young assistants and the other that is successful by letting the disciples walking on himself!". Bazin is the in-controvertible proof that Darcy was of the second kind.

Conclusions

Among many outstanding contributions to hydraulic engineering, his experimental investigation of pipe flow is one of Henry Darcy's richest gifts. His flow resistance formula for cast iron conduits is still in use and the 1858 paper clearly demonstrates that Darcy knew the internal features of pipe flow. The review committee of *Académie des Sciences,* in assessing Darcy's work came to a final judgement that can be assumed as the emblem of Darcy's value. "From detailed analysis of the important work of Darcy, it is evident that he has greatly increased the knowledge that science of engineering received by its forefathers .... *We think that such a research, that needs simultaneously care, perseverance and talent, deserves the highest approval of the Academy........*".

Acknowledgements

The writer offers sincere thanks to his friend and colleague, Professor Willi H. Hager (V.A.W. – Swiss Federal Institute of Technology), without whose invaluable contribution the writer could not have completed the present paper.

References

Bazin, H. (1865). Recherches expérimentales sur l'écoulement de l'eau dans les canaux découverts. *Mémoires présentés par divers savants à l'Académie des Sciences de l'Institut Impérial de France* 19, Paris.

Brown, G., Rogers, J.R., Garbrecht, J. (2000). Task committee planning Darcy memorial symposium on history of hydraulics. *Journal of Hydraulic Engineering* 126(11): 799-801.

Brown, G. (2001). Darcy and the Pitot tube. In *International Engineering History and Heritage*, Proceedings of the Third National Congress on Civil Engineering History and Heritage, edited by J. Rogers and A. Fredrich, ASCE, Reston, VA: 360-366.

Brown, G. (2002). The history of the Darcy-Weisbach equation for pipe flow resistance. In *Environmental and Water Resources History*, A. Fredrich, and J. Rogers eds., ASCE, Reston, VA, pg. 34-43.

Combes, C. (1857). Rapport sur une Note de M. H. Darcy, sur des modifications apportées au tube de Pitot. *Comptes Rendus de l'Académie des Sciences* Paris 45: 638.

Darcy, H. (1834). *Rapport* à M. le Maire et au conseil municipal de Dijon sur les moyens de fournir l'eau nécessaire à cette ville. Douillier: Dijon.

Darcy, H. (1856). *Les fontaines publiques de la ville de Dijon*. Dalmont: Paris.

Darcy, H. (1858a). Recherches expérimentales relatives au mouvement de l'eau dans les tuyaux. *Mémoires présentés par divers savants à l'Académie des Sciences de l'Institut Impérial de France* Paris 15: 141-403.

Darcy, H. (1858b). Note relative à quelques modifications à introduire dans le tube de Pitot. *Annales des Ponts et Chaussées* 28(1): 351-359.

Darcy, H. (1858c). Note relative à quelques modifications à introduire dans le tube de Pitot. *Mémoires* de l'Académie Impériale des Sciences, Arts et Belles Lettres de Dijon. Deuxième série 4: 159-168.

Darcy, H. and Bazin, H. (1865). *Recherches hydrauliques*, entreprises par M. H. Darcy, Imprimerie Impériale, Paris.

Gisonni, C. (2002). Henry Darcy: French engineer. *La Houille Blanche* 57(4/5):97-102.

Gisonni, C., Hager, W.H. (2003). Henry Darcy's 200th birthday. *XXX IAHR Congress*. Thessaloniki (Greece). To be published.

Hager, W.H., Gisonni, C. (2002). Finding Darcy at Dijon. Journal of Hydraulic Engineering 128(5): 454-459.

Hager, W.H., Gisonni, C. (2003). Henry Bazin - Hydraulician. *Darcy Memorial Symposium on the History of Hydraulics*. World Water & Environmental Resources Congress 2003 (EWRI-ASCE). Philadelphia, Pennsylvania.

Poncelet, J.V., Combes, C., Morin, A. (1854). Rapport sur un Mémoire présenté par M. H. Darcy, inspecteur divisionnaire des Ponts et Chaussées, sur les recherches expérimentales relatives au mouvement des eaux dans les tuyaux. *Comptes Rendus de l'Académie des Sciences,* 38: 1109-1120.

Henry Darcy and the Public Fountains of the City of Dijon

Patricia Bobeck[1]

Abstract

Henry Darcy is best known for his empirical law on fluid flow through porous media that he published as an appendix to his book *Les Fontaines publiques de la ville de Dijon*. Darcy had built a water supply system for Dijon in 1840, and in 1856, shortly before his death, he wrote the book to guide other engineers in constructing similar projects. This recent English translation of Darcy's 650-page book provides previously unavailable information on the context for his experiments on water flow through sand, his approach to solving problems of water distribution, a new field at that time, and his views on the humanitarian value of this project.

The book contains 4 parts and an appendix. Part one is a description of the historical water situation of Dijon and attempts to provide water for the city. Part two discusses the construction of the aqueduct and the internal distribution system. Part three presents experiments that Darcy conducted on the aqueduct and distribution system. Part four discusses the expropriation of the springs, which belonged to a nearby village, and purchase of the property under which the aqueduct would be built. The appendix contains eight sections on such topics as the water supply systems of London and major French cities, artificial and natural filtration of river water, Darcy's Law, and pipe making. A separate 28-plate atlas includes drawings of the components of the Dijon water supply system, the Pitot tube, and the apparatus Darcy used for his experiments on water flow though sand.

Introduction

Henry Darcy (1803-1858) wrote *The Public Fountains of the City of Dijon* as a guide for engineers involved in the construction of water supply systems. He wrote the book in the 1850s, long after he had built Dijon's water supply, but before water supply systems were common in European cities.

In planning the Dijon water supply system in the 1830s, Darcy investigated the sources of water available to the city, estimated the city's water needs, and chose an abundant spring located in an adjacent village. Darcy designed and built a 12 km aqueduct and two reservoirs in Dijon. Darcy designed and built a novel (for its time) network-type internal distribution system for the city that allowed a part of the system

---

[1] Geotechnical Translations, 1601 Barn Swallow Drive, Austin, Texas 78746-7430; 512-732-2075; pbobeck@texas.net

to be shut down for repairs while allowing other parts of the city to continue to receive water.

The goal of the Dijon distribution system was to supply water to street fountains for domestic purposes, street washing and fire fighting. At the completion of the project, Dijon became second only to Rome in terms of water quality and quantity. Within the walls of Dijon, street fountains were no farther than 100 meters apart, meaning that no one had to walk more than 50 meters to obtain water, a standard not reached in many parts of the world even today, 150 years after Darcy's death. The pure spring water from the street fountains was free.

Because Darcy wrote the book as a guide for engineers, he discusses a number of topics unrelated to the Dijon project. As a result the book is an encyclopedia of mid 19th century knowledge and technology. He discusses rivers, ponds and lakes as water supply sources; the current understanding of artesian wells; pipes and pipe making; and natural and artificial filtration of river water, among other topics. At that time, sand filters were typically used to filter river water. Darcy developed the law of fluid flow through porous media while studying how to make sand filters smaller, because at that time they were so large that it was difficult to find space to build them.

Darcy's concern for the poor shows in his discussions of the importance of numerous street fountains located so that "the most numerous class can find them easily along their path and are not put off by the length of the journey they have to make" to obtain water.

Darcy's father died when he was about 14, leaving the family in difficult straits. Henry and his brother were excellent students, and their mother "put forth great effort" with the city of Dijon to obtain money to educate them. During his childhood Darcy had been sickened by the only water available and had promised himself to put an end to this situation if ever he were in a position to do so (Paul Darcy, 1957). After Henry completed his education and entered the Corps of Bridges and Roads, the city of Dijon requested that he be assigned to his native city. Shortly after returning to Dijon in 1827, Darcy began working on his plan, and by 1832 he was gauging the Rosoir Spring, which he would divert to Dijon.

The book is divided into four parts and an appendix. The four parts are further divided into chapters. The book is accompanied by 28 plate atlas.

Part One

Part one is a description of the historical water situation of Dijon. Like any engineer beginning a water supply project, Darcy conducted research on previous attempts to provide water for the city.

Chapter 1 is an account of the Darcy's research on old fountains in the city. He acknowledges the assistance of the city archivist who helped him look back through 450 years of city records to determine which springs had been conveyed to Dijon over the years via hollow-log aqueducts, the discharge of these springs, and the location of fountains that dispensed the spring water.

In chapter 2, Darcy examines the Suzon Torrent, a stream that flows through Dijon, as a water source. In 1830, the Suzon was a seasonal stream fed by snowmelt. Darcy investigated the widespread belief that prior to 1830 it had been a perennial stream. Since the 1400s numerous field trips had been conducted along the Suzon to determine why the water disappeared into the streambed, reaching Dijon only during the spring thaw. Many plans had been submitted to remedy the situation, but none had been implemented. Springs located along the river, including the Rosoir, were likewise absorbed by the dry streambed. Darcy investigated the popular belief that the springs had once been more abundant and had made the stream perennial. Based on a review of archives, Darcy concluded that the Suzon had been an ephemeral stream for at least 450 years.

However, the Suzon also presented a more difficult problem. Its streambed, which passed through the city, was a convenient refuse dump for many city dwellers. One of the goals of Darcy's water distribution system was to construct a cover over the sewer and flush it out. In a footnote to chapter 3, Darcy relates that the need to clean up the Suzon sewer was a recurring theme in the historic deliberations of the City Council. Dijon had often been decimated by plague and during these calamitous times all thoughts of the inhabitants had focused with terror on the filth of this sewer, according to Darcy. He says, "It must be agreed that the sanitary measures undertaken by the magistrates in what are called *the good old days* have been almost as frightening as the scourge itself." A parliamentary decree dated 1576 authorized two harquebusiers to enforce the confinement of anyone with plague to cabins outside the city walls. When it was learned that the harquebusiers had refused to shoot the disobedient, the city hired a high executioner to shoot and kill the disobedient ones on the spot. Shortly thereafter a vine grower from a nearby town who had violated the order was tied to a post near the horse cemetery . . . and shot with a harquebus. Darcy says:

"In 1628 the ordinance was renewed, and again in 1630, 1631, 1632, 1634, etc. These savage regulations are very different from the devotion shown today by the clergy, medical doctors and the inhabitants of this department during the periodic visits of cholera. We no longer use the harquebusier to destroy the poison by killing the patient. We simply expose ourselves to dying with him to ease his pain. It is less prudent: it is more Christian."

Darcy accomplished the sanitization of the Suzon sewer in 1847, seven years after the completion of the aqueduct. Darcy discusses this project in more detail in Part Four.

Part one chapter 3 is entitled The Rosoir Spring, the name of the spring Darcy diverted to Dijon. This long chapter contains numerous topics. Darcy begins with a discussion of Dijon's water supply in the 1830s and the fact that drinking water came from private wells and wells along the city streets that tapped into alluvium saturated by fluid from adjacent permeable-wall cesspools. Darcy discusses the mid19^{th} century view of the relative purity of well water, cistern water, pond (stagnant) water, river water and spring water, concluding that spring water was preferable. He provides a chemical analysis of water collected from a well in a private house and a water analysis of the Rosoir water from 1850. Darcy enumerates the potential sources of water for Dijon, which included several springs, a nearby river, and an artesian well. He calculates the amount of water needed for a city water supply, including domestic needs, manufacturing, public buildings, fire suppression, street cleaning, public fountains and gardens. Comparing the numbers with those calculated by English water engineers, Darcy calculates the number to be 150 liters per capita per day, or 4.5 million liters per day for the city. He then eliminates various sources because they do not provide enough water, or they provide water that is too hot in summer, or water that is too expensive because it must be raised from a riverbed. Darcy shows that the Rosoir Spring provides abundant pure cool water that remains cool until it reaches the street fountains because of the underground aqueduct. Darcy discusses water quality in terms of what water should contain: atmospheric air, carbonic acid, sodium chloride and calcium carbonate, and it should dissolve soap well. He discusses iodine, goiter and cretinism. He describes the gauging of the Rosoir Spring in 1832-33. The Rosoir Spring flows from Jurassic limestone and discharges 4,000 to 12,000 liters per minute depending on the season.

In a section on the origin of springs Darcy notes that by 1850 most people believed springs to be fed by infiltrated rain water, and discusses historical views on the origin of springs, including Descartes' idea of underground fires and huge stills beneath the earth's surface. Darcy provides a classification of springs and discusses the history of springs seekers, from the Greeks to his French contemporaries. He describes methods of creating artificial springs.

A significant portion of chapter 3 is dedicated to artesian wells. An artesian well had been dug in Dijon in the early 1830s, but it didn't flow above the ground surface and didn't provide enough water to supply the city, so Darcy rejected it as a water supply source. Evidently during Darcy's time, it was believed that water circulated in pipe-like voids under the earth's surface. Darcy also recognized that many artesian wells flow from sandy layers under an impermeable layer. Darcy understood that friction resulting from water movement consumes hydrostatic head, or pressure, as he calls it. Darcy understood that it was possible to increase the discharge of an artesian well by lowering its discharge point, and calculated that it was also possible to increase its discharge by increasing the diameter of the well. He also understood that as the number of artesian wells from a single source increases, the discharge of the wells decrease. He discusses the Grenelle Well of Paris and the artesian wells of the city of Tours in detail. One drawing in the atlas illustrates Darcy's understanding of artesian wells.

Part Two

Part two describes the construction of the aqueduct and the internal distribution system. In chapter 1, Darcy describes the masonry aqueduct he built between the Rosoir Spring and Dijon. Construction of the aqueduct began in March 1839 and was completed in September 1840. The aqueduct is 12.7 km long between the pavilion that covers the spring and the Porte Guillaume reservoir in Dijon (Figure. 1). For the most part, the aqueduct is 0.60 m wide and 0.90 m high, and is covered by one meter of soil. Manholes are located every 100 m. The spring is located on the bank of the Suzon stream upstream of Dijon, and the aqueduct crosses the Suzon stream 3 times on its path to Dijon. The aqueduct also passes through three villages where water is diverted from the Dijon aqueduct to provide for the villages. Just before reaching Dijon, the aqueduct becomes a viaduct to cross an area of low topography. In the text, Darcy provides details about the slope and cross section of the aqueduct and

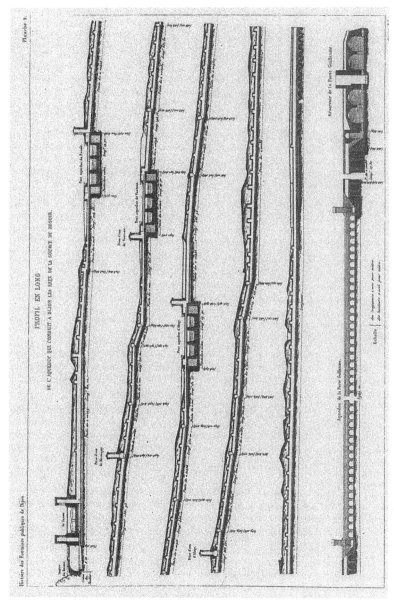

Figure 1. Dijon aqueduct. (Plate 2, Darcy 1856)

details of all the work involved and all costs for material and labor. Six plates in the atlas show Darcy's drawings of the aqueduct project.

In chapter 2, Darcy discusses the internal distribution system, which consists of two reservoirs and a system of cast iron pipes that branch out into all neighborhoods of the city to provide water to street fountains and private concessions. The main artery connects the two reservoirs, distributors branch off the main artery, and service pipes branch perpendicularly off the distributors to serve the street fountains. Darcy used masonry tunnels and trenches for the pipes, depending on the importance of the street above the pipe. The pipes connect at distribution tanks to form a grid system.

The Porte-Guillaume reservoir (Figure 2) is located at the end of the aqueduct and the beginning of the main artery of the internal distribution system. Darcy wanted to be able to do the following at this location: prevent all communication between the aqueduct and the city water supply; prevent water from entering the reservoir without interrupting distribution to the city; stop distribution to the city without interrupting the filling of the reservoir; supply the city using reservoir water only; supply the city with aqueduct flow and reservoir water at the same time; and maintain water pressure in decorative and street fountains at a constant level, the maximum level possible, even when the reservoir was being emptied for repair. He describes the piping system he devised to accomplish these objectives. He brought water into the Porte Guillaume reservoir through a vertical pipe in the central well of the reservoir. At a certain height, the water flowed through some openings and down a stairwell to fill the reservoir. This height was the head that controlled water flow and pressure and maintained maximum water pressure throughout the city; the water supply system has no pumps. Darcy also discusses the problem of air expansion in the reservoir, and the means he devised to minimize this problem.

The Porte-Guillaume reservoir is circular in shape, and is covered by one meter of soil. The reservoir capacity is 2313 m^3. Reducing the water supply to the amount strictly necessary for the inhabitants, which Darcy estimated at 20 liters per person, or 540 m^3 per day, the reservoir could supply enough water for four or five days. Darcy constructed an aedicule on top of the reservoir, which is still there today.

Darcy knew that more storage capacity was required, primarily because an aqueduct repair could easily last more than four or five days. He built a second reservoir, rather than one larger reservoir, to prepare for the possibility that the main artery would require repair. By building the Montmusard Reservoir on a hill at the other end of the main artery, Darcy ensured that all points in the city could be supplied by interconnecting pipes. In addition, Darcy saw that supplying the main artery from both ends made it possible to furnish a much larger amount of water than if it were served by only one end. Another reason Darcy gives for constructing the second reservoir is the possibility that the Porte Guillaume reservoir may need repair. In that case, the springs themselves could not supply enough water for street watering during the hot summer when domestic water use would be high, but with the water from the Montmusard Reservoir, it would be possible to do so.

The Montmusard Reservoir is rectangular and is located underground, covered by one meter of soil. Its capacity is 3177 m^3. All flow in the city pipes was suspended every night while the Montmusard Reservoir was being filled.

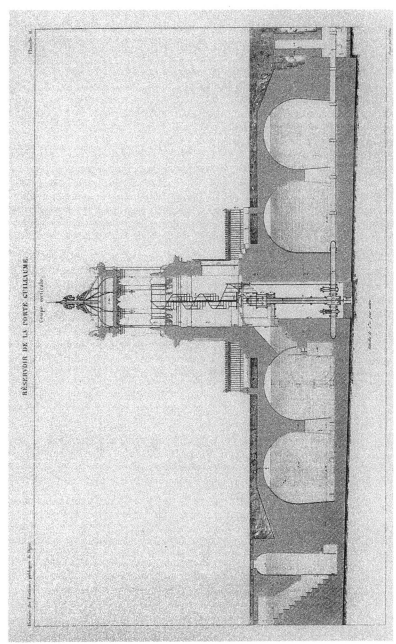

Figure 2. Porte Guillaume Reservoir. (Plate 11, Darcy, 1956)

Chapter 2 also contains a detailed discussion of each of the ten major distributors, whether they were installed in a tunnel or trench, the connector pipes located along them to connect them to another distributor, the distribution tanks located along each distributor, the street fountains located along each distributor and their locations. Plate 8 of the atlas is a city map that shows the locations of the main artery, distributor pipes and street fountains. Darcy describes the valves, drain valves, air release valves, and distribution tanks used in the system. He also discusses the types of pipes used in the distribution system and the methods used to connect them. He provides the calculations for the jets of the fountain he built at the Place Saint-Pierre, which is still there today.

He provides a detailed description of the street fountains, which are a major component of the distribution system because they provide water for domestic purposes, street cleaning and fire fighting. Street cleaning was necessary because of the use of horses for transportation and also because open sewers ran either down the center of the street or along the edges of the streets. Darcy maintains that supplying water for domestic purposes is more important than street washing. He quotes Mr. Emmery, Chief Water Engineer of Paris:

> "The experts will add that it is especially necessary to give the unfortunate class the possibility of doing more of all kinds of washing, including their bodies and the clothing that is distributed so sparingly to each individual. The experts will repeat to you that this is how you can effectively attack the problem of improving the sanitation of a large city. Such is the immense service that drawing water from street fountains can provide."

Darcy writes that Chief Engineer Emmery considers the lack of a street fountain as a real calamity because it causes an increase in mortality for the unfortunate class. According to Darcy, "the City of Dijon seems to have been inspired by Mr. Emmery's thinking. The City took up this plan and all its consequences. The City of Dijon did good almost to excess, if there can be excess in good."

The plumbing in the street fountains had to allow for on-demand and constant flow to satisfy these three requirements. The street fountains also required sewers to drain away water to prevent them from freezing in the winter. Since Dijon lacked a storm or sanitary sewer system, Darcy had to devise methods to solve this drainage problem. Chapter 2 ends with a detailed accounting of the costs involved in the construction of the internal distribution system. Twelve plates of the atlas contain Darcy's drawings of the components of the internal distribution system.

Part Three

Part three presents experiments that Darcy conducted on the aqueduct and distribution system. Chapter 1 discusses experiments on water flow in the aqueduct that conveys water from the Rosoir spring to Dijon. These experiments are primarily on velocity in various parts of the aqueduct, straight portions vs. curved portions, portions with smooth walls and portions that include falls.

Chapter 2 deals with experiments on water flow in the conduit system. Darcy had long been interested in studying how engineers could accurately predict a pipe's discharge. At that time, engineers depended on the formula de Prony had proposed on the basis of a few experiments, but the discharges calculated by the formula were seldom duplicated in practice, with the result that engineers could not predict how their systems would perform. Darcy had encountered similar problems in the construction of the Dijon water supply system. He wanted to conduct new experiments to study this problem, but in Dijon he did not have the equipment he needed (Paul Darcy, 1957). The opportunity arose from what was undoubtedly a painful episode in Darcy's life. The revolutionary government that took power during the events of February 1848 stripped Darcy of his post in Dijon. The people of Dijon protested, as did the Municipal Council, the Corps of Bridges and Roads and the Ecole Polytechnique. The revolutionary government realized it had made a blunder and assigned him to work on a canal in a neighboring village where he stayed three months. After elections of May of that year restored some semblance of order, Darcy was named Chief of Municipal Service in Paris. In 1849 he was promoted to Inspector General First Class. In 1853, after Darcy had worked as consultant on the Brussels water supply system, the City of Paris made a small factory available to him, the Minister of Public Works provided him a grant and assigned an engineer as his assistant. At last Darcy was able to repeat de Prony's experiments and conduct new experiments with the abundant equipment at his disposal. In his experiments, Darcy showed that the smoothness or roughness of the inside wall of the pipe had a great influence on the pipe's discharge. This memoir was published as *Recherches experimentales relatives au mouvement de l'eau dans les tuyaux [Experiments on Water Movement in Pipes]* in 1857. The first three sections of chapter 2 summarize this memoir, and one plate in the atlas illustrates the discussion.

Darcy then presents the results of a study on the Dijon pipe network in which he shows that the coefficient of resistance in the main artery increased between the 1840s and the 1850s. In this chapter he also discusses the influence of air resistance on the height of water jets, the way in which a second reservoir can increase pipe discharge, and principles to guide an engineer in calculating pipe diameters for a water supply system.

Part Four

Part four contains a discussion of the administrative and judicial questions involved in the construction of the water supply system. These included the expropriation of the spring, the opposition mounted by the owners of mills located downstream of the springs, the purchase of land the aqueduct would cross, the sanitizing of the Suzon sewer within the city, and private water concessions.

Because the Rosoir spring was located within the jurisdiction of a nearby village, the question of water rights had to be resolved. The question was settled in favor of the City of Dijon by edict of the King on 31 December 1837. The city compensated the state and the village for the spring. To prevent "rash claims" Darcy proposed a water distribution formula that gave the inhabitants of the villages between the springs and Dijon 150% of the per capita water allocation of a Dijon resident. The mechanisms that divide the water are described in part two chapter 1. The claims of the mill owners were finally resolved by the payment of simple damages rather than expropriation costs.

At that time juries were used to establish expropriation costs, and Darcy dreaded their intervention in the process of purchasing land for the aqueduct. He says,

"It must be remembered that at this time, juries often granted sums that even the most greedy landowners would not have dreamed of. But I was determined to confront this little storm by counting on the good sense of the population if I was able to reduce the leaders to silence."

He did so by collaborating with the mayor to draft a deed of sale to purchase subsurface rights for the aqueduct, which would be buried at a depth of one meter, while preserving the landowner's right to cultivate the surface and to collect the fruits of it without paying any rent or fee to the city, as long as they did not do any digging, construction or planting of vineyards and would allow the city to construct markers indicating the presence of the aqueduct and allow the city access to maintain the aqueduct. Darcy enlisted "an expert" to come to the town hall of each village to meet with the largest number of owners possible to discuss all the conditions of the contract with them and to invite them to themselves set the indemnity to which they thought they were entitled. Darcy says of this process:

"I have noticed after addressing numerous meetings that it generally happens that greedy emotions are neutralized. It seems that bad instincts dare not show themselves openly. It happened as I expected. The discussions were conducted with so much calm and sincerity on the part of the expert that the landowners exhibited such rare moderation that I was obligated to increase the amount of the indemnities claimed by one of the communes. From that time forward, so many landowners felt confident of us that they sent us the acts of sale, signed blank. Not a single opposition was raised, and there were five hundred fifty-six parcels."

The flushing of the Suzon "sewer" and the removal of the immense accumulation of rubbish and decomposing animal and plant matter that in places formed a layer 1 to 2 meters thick was not accomplished until 1847 because of the opposition by riparian residents who objected to the city ordinance requiring them to remove outhouses and sewers that they had built there without permission. A judgment by the Civil Court in 1846 ended the legal challenges, and Darcy says, "all the opposing parties thus had to yield in the presence of the text of these decisions and today they no longer have any trace of the bewilderment that misled them into their resistance." The Suzon streambed was enclosed in an aqueduct. As a result, Darcy says:

"This city no longer has to dread the deadly effects of emanations from the sewer and owners of property located 100 to 150 meters from the banks also gained as a result of the enclosure that now protects them from periodic flooding of their cellars and from contamination of their well water by infiltration."

In 1844, Dijon residents could purchase water concessions to pipe water directly into their homes. As a first principle, Dijon did not allow free water concessions for anyone, because it was well known that this practice was greatly abused in Paris. Darcy states, "institutions that are created with taxes paid by all citizens must be used for the benefit of all." Apparently few Dijon residents felt the need for private water concessions, because in 1855 at a time when Dijon evidently sought to increase revenue from the water supply system Darcy wrote to the mayor of Dijon expressing surprise at the low number of private water concessions. Darcy attributes this to the abundance of street fountains and the high cost of connecting a private concession. Darcy suggests policies that would not be costly to the city that would encourage more private water concessions. In the process, he reiterates his position that the city continue to supply free water through the street fountains and that "Water must not be restricted for [the less privileged class] any more than light and air are restricted for them."

Appendices

The Appendix contains eight notes, designated by the letters A through H. Appendix A is a list of springs located near Dijon. Darcy probably made this inventory as a part of the process of selecting a water supply source. Appendix B is a contract dated 6 December 1445 between the City of Dijon and Pierre Belle, a carpenter from the neighboring village of Talant. In this contract, the carpenter agrees to bring the waters of the Montmusard Spring to the Porte Saint-Nicolas, one of the old city gates of Dijon through an aqueduct constructed of hollowed-out logs. Darcy included it because it was the first deliberation of the city on the establishment of fountains in Dijon, and, as Darcy says in footnote 4 to part one chapter 1, it is a very curious contract. The contract was included in Darcy's book in the original Old French.

Appendix C is a discussion of the water supply systems of London, Paris, Brussels, Lyon, Bordeaux, Nantes, Besançon and Nîmes in the 1800s. Darcy's research on what other water supplies is an activity that engineers today would conduct prior to designing something new. Darcy discusses the sources of the supply, the quality of the water, the number of houses served, the cost of the water and how the water is paid for. At that time, London drew all its supply from rivers, and most of the houses in London were supplied with water. (London did not have street fountains.) Everyone paid for water and the water bill depended on the size of the house. According to Darcy, the rates were high and profits were excessive. In 1850 only 20% of the houses in Paris had water concessions. The water came from rivers. Most Parisians purchased water that was carried to houses by water carriers. Darcy discusses his colleague Belgrand's plans to bring more water to Paris. In this note, Darcy also discusses his adaptation of the Pitot tube to make it easier to use in gauging water flow in rivers and streams. The atlas contains a drawing of the Pitot tube.

Appendix D, entitled "Filtration" contains an account of the experiments that led Darcy to formulate the law we call Darcy's Law. The note begins with a discussion of the fact that the chemical purity of river water is often compromised by turbidity. Because settling ponds don't completely clarify the water, filtration is required. Darcy defines artificial filtration as the process of passing water under variable pressures through beds of sand and natural filtration as the process of passing river water through its own alluvium. Darcy cites London and Glasgow as cities that practice artificial filtration. The disadvantage of artificial filtration is the large surface area required for the filtration beds, 4,000 square meters in the case of one filtration bed in London. Darcy proposes a modification to decrease the size of filtration beds by increasing the discharge of the filter, using a taller column of water or negative pressure under the filter. He describes the natural filtration systems of Toulouse, Lyon and Perth, Scotland. He discusses how the product of a natural filtration gallery is affected by changing its length, width and depth. This leads Darcy into the discussion entitled, "Determination of the Laws of Water Flow through Sand" and his description of the experiments he conducted in Dijon in 1854 with Engineer Ritter. Darcy completes this note with more ideas on springs and a discussion of a book entitled the Art of Finding Springs, by his contemporary Mr. Paramelle. Three plates of the atlas are Darcy's drawings of artificial and natural filtration and an artificial filter he designed. Figure 3 shows natural filtration galleries and the equipment Darcy used in his experiments on water flow through sand.

In Appendix E, Darcy discusses the methods he used to gauge the Rosoir Spring. In Appendix F, Darcy discusses methods for drawing a constant volume of water from a variable level channel. Darcy's drawing of this equipment is shown on one plate of the atlas. Appendix G is a discussion of pipe strength and fabrication of cast iron, lead, sheet metal and bitumen pipes. Sheet metal pipes covered with bitumen were a new invention in the 1850s. Two plates of the atlas show drawings of pipe-making equipment. Appendix H contains additional information on water flow in the Rosoir Aqueduct.

Conclusion

In this brief overview of Darcy's book, I have had to leave out numerous topics that Darcy discusses. The reader of Darcy's entire book will encounter many treasures not mentioned in this article, including a discussion of the cisterns of Constantinople and an ingenious fire-suppression plan for Dijon's Theater that uses the hydraulic head of the water supply system to replace many hours of men's labor to raise water to the roof of the Theater. I have tried to provide insight into the personality of this modest, civic-minded engineer who grew up during Napoleon's Empire and produced his great accomplishments during the alternating empires, constitutional monarchies and republics of 19th century France.

The people of Dijon are grateful not only for the water supply system he built but also for his work to ensure that the Paris-Lyon railroad route built in the 1840s would pass through Dijon. He surveyed the route, presented it and obtained the approval of the national government, and he himself supervised the construction of a 4.1 km tunnel through the mountains near Dijon. The railroad brought Dijon increased population and prosperity. In addition, Darcy was a city councilor, a founder and administrator of the Society of Mutual Help, administrator of the Savings Bank and the Hospice, and a founder and promoter of the Cooperative Production Societies, all organizations that sought to lend a hand to those like him who had had difficult beginnings (Paul Darcy, 1957). At the time of his death, the population of Dijon wanted to send him to Paris as their Deputy, roughly the equivalent of a senator in the U.S. political system. After his unexpected death in Paris, his body was brought back to Dijon by train, and the entire city gathered at the train station to show

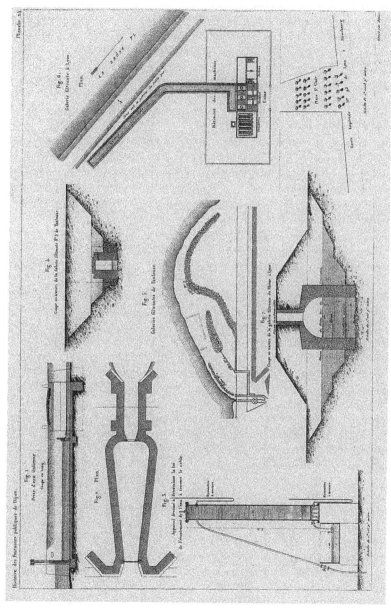

Figure 3. Natural filters and column apparatus. (Plate 24, Darcy, 1956)

their respects. Shortly thereafter, the City of Dijon changed the name of the Place du Chateau d'Eau [Water Tower Square], the location of the Porte Guillaume Reservoir, to Place Darcy in his honor.

References

Darcy, H. (1856). *Les Fontaines Publiques de la Ville de Dijon.* Dalmont, Paris. 650 pg., 28 plates.

Darcy, P. (1957). *Henry Darcy: Inspecteur général des ponts et chaussées, 1803-1858.* Imprimerie Darantière, Dijon. 63 p. Unpublished English translation by Patricia Bobeck.

Henry Darcy - Biography By Caudemberg

Willi H. Hager[1]

Abstract

Henry Darcy (1803-1858), engineer, designer of Dijon's water system, renowned researcher who provided the basis for both Darcy's Law and the Darcy-Weisbach equation, and initiator of the open channel research program completed by Henry Bazin (1829-1917), is being commemorated on the bi-centennial of his birth. This paper reviews his abilities by providing an original English translation of an obituary by Girard de Caudemberg, an intimate friend and colleague of Darcy's. Historians and engineers have largely overlooked Caudemberg's obituary. The story of the gold medal presented to Darcy for his services to the city of Dijon is also reported here, as well as personal observations.

Figure 1. Portrait of Henry Darcy by F. Perrodin,
from the Collection of the Bibliothèque Municipale de Dijon.

---

[1] Professor., VAW, ETH-Zentrum, CH-8092 Zurich, Switzerland

Introduction

Henry Darcy (1803-1858) ranks among the most important hydraulic engineers because of his outstanding services in both civil engineering and hydraulic research (Figure 1). The hydraulic community, especially that of the United States, has proposed a special symposium to commemorate the bicentennial of Darcy's birth in 2003. Another festivity is proposed at Dijon, natal city of Darcy and the place where he practically spent all his life. Dijon has largely profited from Darcy's presence, particularly because of the drastic improvement of drinking water quality, the design of an appropriate sewer system, and maybe most important his ability to bring the railroad and thus prosperity to his hometown.

Hydraulicians are certainly familiar with his name, since we use Darcy's law for groundwater flow, the Darcy-Weisbach equation for closed conduit flows, and the groundbreaking results presented by Darcy and Bazin relating to open channel flows. The historically interested engineer may have come across the 'official' obituary of Darcy written by Charié-Marsaines (1858). However, there exists a larger note by a close friend to Darcy, Scaevola Charles Girard de Caudemberg (1793-1858): Who, like Darcy, was an engineer in the Corps des Ponts et Chaussées, but with a more applied engineering career than his friend. His note was published in a journal that is challenging to locate, the *Mémoires de l'Académie Impériale des Sciences, Arts et Belles-Lettres de Dijon*. Volume 4 of Series 2 (1858) contains the remarkable paper of a close friend to Darcy, written in a style that is currently out of fashion, but with much intimacy, great love and an admiration for a person to whom one would wish only the best. The paper's title page is shown in Figure 2. Sadly, Caudemberg passed away the same year as Darcy, under conditions that are unknown to the present author.

It was thought that a translation of the Caudemberg obituary and its appendices would be a worthwhile contribution to the Darcy symposium, and of interest to technical historians and the engineering community. In addition, an abridged translation of a paper by Petit (1987) is presented in the Appendix that describes the steps taken in creating the memorial medal to Darcy in 1846. Italicized text are original French titles, text in curly brackets are footnotes from the original document, while text in square brackets are translator comments.

Of final note, mid 19th century French writing is elegant and has a complex sentence structure. Caudemberg was clearly comfortable with the style and put good effort into his work. Thus, most passages lack the concise character of modern technical writing, are difficult to interpret directly, and any translation must make compromises. Nevertheless, this translation has attempted to maintain the original document's tone, tempo and quality. Hopefully this will allow the reader to feel the depth of emotion felt by Caudemberg.

Caudemberg Biography

[Introduction]. "Gentlemen, in honoring me as its choice to talk about Mr. Henri Darcy, our dearly beloved and knowledgeable member whose recent loss we deplore, the Academy has anticipated my heartfelt wish. Nothing could soften the bitterness of such a cruel separation and the sharp and unanticipated grief than to speak to you at length and in detail about this friend whose rare and excellent

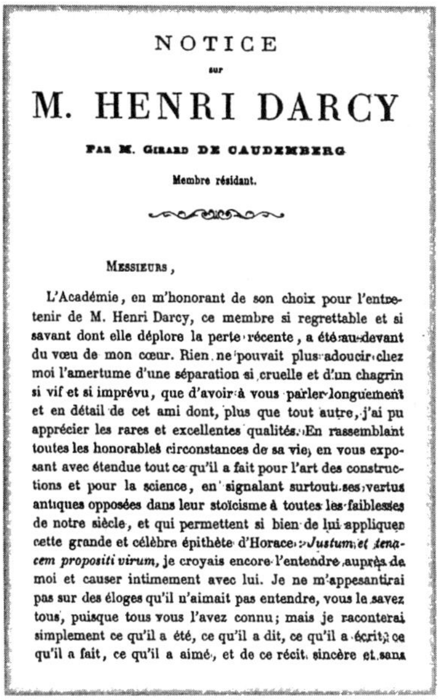
NOTICE

sur

M. HENRI DARCY

PAR M. GIRARD DE CAUDEMBERG

Membre résidant.

MESSIEURS,

L'Académie, en m'honorant de son choix pour l'entretenir de M. Henri Darcy, ce membre si regrettable et si savant dont elle déplore la perte récente, a été au-devant du vœu de mon cœur. Rien ne pouvait plus adoucir chez moi l'amertume d'une séparation si cruelle et d'un chagrin si vif et si imprévu, que d'avoir à vous parler longuement et en détail de cet ami dont, plus que tout autre, j'ai pu apprécier les rares et excellentes qualités. En rassemblant toutes les honorables circonstances de sa vie, en vous exposant avec étendue tout ce qu'il a fait pour l'art des constructions et pour la science, en signalant surtout ses vertus antiques opposées dans leur stoïcisme à toutes les faiblesses de notre siècle, et qui permettent si bien de lui appliquer cette grande et célèbre épithète d'Horace : *Justum et tenacem propositi virum*, je croyais encore l'entendre auprès de moi et causer intimement avec lui. Je ne m'appesantirai pas sur des éloges qu'il n'aimait pas entendre, vous le savez tous, puisque tous vous l'avez connu; mais je raconterai simplement ce qu'il a été, ce qu'il a dit, ce qu'il a écrit, ce qu'il a fait, ce qu'il a aimé, et de ce récit sincère et sans

Figure 2. Caudemberg Biography title page.

virtues I perhaps more than any other person have been able to appreciate. While I was considering all the honorable events of his life and discussing at length everything he did for the construction and scientific professions, pointing out in particular his old-fashioned virtues that contrast in their stoicism with all the weaknesses of our century, virtues that allow the application of this famous epithet of Haorace *Justum et tenacem propositi virum*, to him, I thought I could still hear him next to me and I thought I was still chatting intimately with him. I will not dwell on the elegies that he did not like to hear, as you all know, since you all knew him. I will simply tell what he was, what he said, what he wrote, what he did, and what he loved and from this sincere account a completely natural elegy will emerge unadorned and at this time when his ashes are still burning in your memories, this will be more than a notice it will be a funeral oration.

[Early career]. Henri Darcy was born in Dijon on June 10, 1803. At the age of fourteen, he lost his father, and the duties of educating him and his younger brother were entirely in the hands of his mother, who loved the two of them, and who knew, through her tender solitude and her wise counsel, they would find their very different careers, by solid and extended education that would later allow them to obtain the highest positions in public life.

Henri Darcy went to the college preparatory school of Dijon where he showed his talent in mathematics, Ecole Polytechnique being his goal that he entered in 1821, in the first ranks, and advanced in knowledge so much that he was able to gain entrance to the Ecole des Ponts et Chaussées. In 1826, he was assigned to the county of Jura, as an engineer but left when a similar position became vacant in Dijon the following year. That event, fortunate both for him and for us, returned him to his mother, brother, friends, and family, and gave him family life on his native soil, which he was made for, to which he brought that extraordinary high spiritedness that is the charm of close relationships.

For a long time, the city of Dijon had looked for methods to provide good drinking water for its inhabitants, so that they could abandon wells that offered an unhealthy and unpleasant drink. Since the XV century, the spring of Rosoir had gathered attention among the citizens. Others had proposed to use the waters of the Suzon. Still others wanted to pump the waters of the Ouche by machines, or even bring to the city the waters of the Neuvon and Chartreux or other springs.

However, in 1828 a novel solution to that problem presented itself with a chance of being accepted by everyone. The drilled well, so-called Artesian wells had become the fashionable, and the recent remarkable successes in Paris by Mr. Mulot could provide hope for excellent results. It was under the power of these ideas that a society was founded at Dijon to acquire a drilling apparatus to advance a probe. The municipal council supported the effort with an initial amount of 3,000 francs and chose Place Saint Michel for the test. Operations started in March 1829, under the direction of the departmental engineer, with Mr. Darcy taking part. On August 6, 1830, the probe had advanced at 150.72 m below ground and entered suddenly into a void or reached the groundwater table, and water discharged up the tubes, but did not spring into the air. The work that continued from that time to increase discharge stabilized the water level at 2 m below the Place. Mr. Darcy noted that the spring produced water of excellent quality but that even with lowering the level to 10 meters below the Place pavement, it could provide only 500 liters per minute, a quantity insufficient for the city of Dijon, even if that quantity had been free flowing. But it was not, and it was impossible to hope for water at other sites. Mr. Darcy thus reported on all his hopes and his laborious researches to supply waters to his hometown by the surrounding springs.

[Rosoir Spring]. The Rosoir had impressed him in early investigations as fulfilling all the desirable conditions, and he started to study the spring with the wisdom that he demonstrated through his entire career. He first determined the quantity of water that would satisfy the various needs of the citizens and the city's sanitation for cleaning roads and rivers. Based on the numbers available from various cities of England and France, he was able to establish with a high degree of probability an amount of 150 liters per day per capita for both private and public use. That would require 4,500,000 liters per day for the 30,000 inhabitants, or 3,125 liters per minute.

Once that criterion was selected, he had to ensure that the Rosoir spring was able to supply that volume, even through the driest months and the years of the least rainfall. Mr. Darcy investigated that point for several years with hydraulic measurements, the details of which are contained in the very remarkable book *Les Fontaines Publiques de la Ville de Dijon*, published in 1856, that you keep in your libraries and to what we will come back later. The results of these experiments gave a minimum spring discharge of 2,770 liters per minute. Therefore a deficit would result for the required supply, in addition to 255 liters [per minute] needed for the communities of Messigny, Vantoux and Ahuy. It is necessary to add that this deficit would occur exactly in the period when the maximum discharge is needed.

Appropriate engineering solved the problem. The Rosoir spring was located on the right side of Suzon, but the left side of the river was better suited for the aqueduct. To satisfy this condition, the entrance to the aqueduct was a small tunnel below the river allowing the water to flow to the left side. The tunnel invert elevation was set at 1.1 m below the riverbed, and Mr. Darcy saw in this operation a guarantee to increase discharge during the dry season. That was confirmed by later precise measurements made once the works were completed. A minimum of 4,200 liters was supplied, thus the discharge was increased by one half.

Once that important problem had been solved, the able engineer asked for still more. He wanted his citizens to always have very pure and fresh waters even in the summer. The analysis of the water, made with great care by our chemist Mr. Sainte-Claire-Deville, produced only 2,607 mg of solid matter for 10 liters of water. Accordingly, it is one of the purest fluids that can be imagined, containing iodine in sufficient quantity to avoid problems with cretinism or the goiter endemic, if one can indeed assign those degrading affections to the absence of that element. Like all large springs, the Rosoir, remains at a constant temperature in summer and winter at close to 10°C. However, in most constructed water mains, little care is taken to preserve the temperature to the delivery point where it is used such that the inhabitants have cool drink in the summer. Mr. Darcy wanted everyone to find cold water just in front of his door to quench his or her thirst, even during the hottest time of the year. That can only be obtained in Paris by using ice or deep-well water. Burying the aqueduct or the pipes downstream of the spring by a minimum of 1 m below ground was sufficient, but it resulted in variable slopes and even drops in the aqueduct profile. Observations made since the construction of the aqueduct, between Porte Guillaume and the other main fountains proved the success of this simple method. In both heat waves and cold weather the water temperature has not varied more than 2°C from that of the Rosoir spring.

With a convenient mechanism, the faucet of every fountain pours water with profusion for all that come to look for it. Accordingly, Mr. Darcy not only supplied the city with abundant water quantities throughout the year, but he made also all that it was distributed to all the population, whether poor or rich, in the best conditions possible, and absolutely free.

While promoting this system of generosity in the councils of the city, he placed the praise to the advantage of his forecasts. Such is the true justification of the recognition that is due to him in perpetuity by his country, where one enjoys, where one will enjoy, these exceptional benefits for a long time, without perhaps thinking that nowhere elsewhere, even in Rome the city of the fountains, they were not granted to the population. Mr. Darcy's disinterestedness was proved by his refusal of the 50,000 francs commission that was lawfully his for the beautiful work, which in spite of the fortune that an advantageous marriage had given him, was far from was can be disdained. However, in our sense, sirs, these were the valid rights that he had acquired largely for the perfection of his work. [It rightfully added to his recognition.]

The best things are often hindered by shabby quibbling and special interests with little vision that raise nearly insurmountable obstacles. Gentlemen, you all know that without the energetic support that Mr. Darcy found in the municipal and prefecture authorities, the execution of the project would maybe still be waiting. {The names of the mayor, Mr. Dumay, and prefect, Mr. Chaper, are in all memoirs. It appeared useless to mention them in a text that must be dedicated to a unique memory and recent regrets.}

Finally, when all the difficulties were solved, the works on the diversion of Rosoir started on March 21, 1839. On August 1, 1840, the city bought the spring and the land needed by expropriation. Water arrived the following September 6th at Dijon, after a time of 3 hours, 33 minutes in the masonry aqueduct that was 13,000 m long, with 53 m of slopes and across numerous bends. It was a great day for the city and a triumph for the engineer, when under acclamations of the population; the first drops of water noisily entered the magnificent reservoir at Porte Guillaume. On May 29, 1839, Mr. Darcy, then an ordinary engineer, was promoted to acting chief engineer of the department. On May 7, 1840, he was permanently promoted to that grade. These merits thus followed the success of his works. Read in the book already cited the details of the many problems solved by Darcy relating to the internal distribution of waters, and the creation of the fountains of the Rosoir, the Saint Bernard quarter and the plaza of Porte Saint Pierre, and if you wish to have an exact idea of the work and the meticulous care with which all the various needs for the city were envisaged and satisfied. The limits of this note do not permit us to approach the immense developments of such a topic; developments that one would not suspect if Mr. Darcy himself had not revealed them.

[Railway Project]. The City of Dijon owed to that restless engineer a service that was even greater than the perfection of fountains. The main railroad from Paris to Lyon was being designed, with the portion of Paris to Montereau and the extension to Troyes already executed. It remained to be decided whether to cross the Seine valley and head towards the Côte d'Or and Dijon, and then on towards the Saône valley, or starting from Montereau, follow the Yonne valley and bypass Dijon. These projects had been studied with a predilection for economic design and resulted in a preference for the Yonne route, in spite of the efforts of Dijon. With this decided, the Yonne and Saône valleys had to be connected directly, or by following the Ouche River. In both cases, the top of the mountains separating the Seine and Rhone catchments had to be tunneled. Two projects were again proposed, one across the Serein valley without directly passing Dijon, the other across Mont-Afrique, also not particularly advantageous for the capital of Burgundy. However, a superior person came to her aid [Darcy]. Mr. Darcy, with an economic background that is important for engineers, proposed another railway alignment following the Brenne and Oze valleys. A tunnel at the Blaisy heights would allow passage through Dijon. He investigated his proposal in detail, defended it in front of the Ponts et Chaussées Council with a conviction and a power of reasoning such that he and thus Dijon finally won. However, another engineer executed what he had designed! {We will not commit the omission here of not citing the devoted and influential Mr. Saunac, then deputy of the Côte d'Or, who is held in great esteem in the Chamber, and whose contribution made for the success of the most efficient utility.}

When you presently look at the large expanse of the Dijon station and its immense traffic you cannot help to acknowledge that Mr. Darcy has rendered a great service to the railway company [PLM]. The company would have bitterly regretted the other projects, because they had not joined a great commercial center and a large population with the cities of Besançon, Belfort and Mulhouse.

On January 15, 1845, once these great services, these good and bold applications of science were completed, Mr. Darcy had earned the right to become a resident member of Dijon Academy. He took part for three years at your meetings and in your works. With rhetorical qualities and the originality of an open and brilliant mind, he took interest in the reports that he was asked to prepare, and the discussions in which he was involved.

[Political Change]. However, a storm grew on the horizon that separated him from you and his hometown, and troubled his existence as others also. The ominous days of February 1848 resulted in a radical revolution in the center of France. In these moments of great political catastrophes, men of heart are always those in highest danger. Because of their conviction and disdain to vainly conceal condescension to the errors of the day, the leaders dread their influence and courage. Despite his eminent titles, those that had seized power prosecuted Mr. Darcy. He was dismissed from his functions and transferred from Côte d'Or, an irreparable circumstance for his feelings, but which in the end turned out to be successful for his further career and fame.

I cannot better characterize that turbulent phase of the career of my unfortunate friend than by repeating the words expressed in March 1850, by the honorable president of our Academy. In responding to Mr. Collin, who replaced Mr. Darcy as resident member, he stated, "When considering all qualities assembled in one man, unselfishness, genius and character, he received after a great political storm the ingratitude of some of us. These letters of ostracism that are already in antiquity, were the price of courage and of outstanding service."

It was at other locations and in other fields with more impact where we may follow our worthy colleague. At his new residence in Bourges he had to solve a problem cited in an *Etats de Berry* memoir, which had waited for a solution since 1780. The problem had gained the attention of the National Assembly in 1790 and was mentioned in 1820 in *Le grand exposé du système de la navigation de la France*. It was reconsidered in 1848 with great and urgent interest because it proved large demands for the unemployed masses following the commercial stagnation. The problem was the drainage and cultivation of the Sologne County. It was a land deserted due to fever and sterility, where the population and animals languished thinly and stunted. That question of national utility was worthy of Mr. Darcy's investigation, who then chaired a commission of the Ponts et Chaussées. In those cases, as we know, all work centers around the chairman, and he was not a person to repudiate such a task. I have under my eyes his draft plan that testifies to a very widespread investigation, in which his colleagues Machard and Brongniard aided. The Sologne has a surface of 460,000 hectares, of which half was used agriculturally and gave a poor harvest due to insufficient manures and amendments. In its west, the soil was dominated by clay that created large swamps, whereas the east had sandy soils where the plants were burnt. Therefore, that land had to be improved to allow for cultivation, for tree plantations and animal production. About 300,000 hectares had to be covered with pine trees that were appropriate for the soil. The canal of the Sauldre that was partially built between the Cher and Loire rivers was referred to as the Grand

Sologne Canal. It provided both water flow and cheap transportation of goods. The railway line crossing the Sologne River between Orléans and Vierzon was also constructed for that purpose. Secondary channels had to be opened and roads were planned across the wastelands to provide the transportation that was lacking in the neglected territory. Strangely, the Sologne had not always been what it was then, with a number of castles demonstrating that there had once been prosperity.

I have quickly given these details to underline the importance and the extent of the works that Mr. Darcy was asked for in a short time. He presented a draft to fix the future of that region that should be consulted when fortune and fertility have to be improved. Before having completed these works, compensation was offered to Mr. Darcy for the unjust treatment that he had previously received. Bourges was an exile for him. He was thus called to Paris and appointed Director of Roads and Water for the capital. This was an acknowledgement of his high merit, and it provided him the opportunity to demonstrate his abilities completely.

[Mission to London]. During his long study of the Dijon aqueducts, Mr. Darcy noted the inadequate equations for flow in conduits by Prony. He profited from a favorable occasion to verify by numerous and extended experiments the coefficients of the Prony formula, and their invariability seemed doubtful. The vast establishment of Chaillot was located within his new service, which contained large reservoirs, considerable chutes that were simple to modify, and all the accessories needed for large-scale experimentation. He did not waste a moment to prepare the execution. That work was hardly started when he was sent to England by the French government to study mainly in London the methods of Macadamization for street construction. The study would accomplish all the elements needed to introduce the system for the principal roads of Paris.

The mission entrusted to Mr. Darcy was urgent. He traveled to England to study all kinds of roads used in London where the system was employed. He thus compared the construction, maintenance and the ease to tow cars for a variety of road designs. The factors that force London to use tarmac, even in the streets of the city, are the narrow lanes with immense traffic. It is the loaded and fast carriage passage, of the omnibuses especially, in the streets of the City of London that caused an unceasing noise. It was intolerable to the inhabitants, and the elevated houses that line the streets underwent an unending shaking. But how to preserve these narrow streets from the mud and the dust, so inconvenient on the pavements of our boulevards?

Mr. Darcy taught us the secret that involves very hard and carefully fractured gravel, either of granite or porphyry, collected at distant locations. The City of Paris was shocked by those expensive proposals, but the immense sums spent for the maintenance of its roads were then significantly reduced. Stones too soft break almost immediately. For example, the maintenance of only one square meter of the boulevard between Madeleine and Bastille Place had a cost four times as high than the previous pavement expenses. The incessant loads almost prevented the movement of heavy cars, and dirt and dust could hardly be avoided. In his paper on the roads, Mr. Darcy reported the circumstances to improve that situation. His paper summarizes all details. Like all the works of that engineer, it shows the seal of a persevering mind that wants to detect all, highlights all questions, inquires all, deepens all, and leaves nothing to those who would like to express their opinion later.

[Chaillot experiments]. After having returned from England, the experiences of Chaillot were pursued with a new zeal. They were not slowed down because Mr. Darcy was appointed Inspecteur divisionnaire on April 30, 1850, a title that was later replaced by the inspector general of second degree, without any financial effects. {Mr. Dupuit, who followed Mr. Darcy in the service of the streets and water of Paris, did not wait for the initiative of the administration to gracefully continue and facilitate his comrade's continued important experiments}. These experiments had the purpose of defining the coefficients in the pipe flow formula for all relevant cases of practice, starting with the smallest diameters to a maximum of 1 m, and for velocities that varied between some centimeters to six meters per second. Darcy determined the coefficients of the general formula with the resistance expressed by the two first powers of velocity, in addition to the formula containing only the quadratic velocity term. That latter expression applies for velocities larger than 1 m per second, whereas the formula with only the first power of velocity may be used for velocities smaller than 10 centimeters per second. The comparison of computation and observation proved that for velocities in excess of some centimeters per second the resistance may be reproduced by the quadratic expression and the effect of the linear term may then be dropped. Mr. Darcy stated that the simple formula applies particularly for flows in used pipes that have a cover of deposits, thus their ordinary state.

By comparing the numerical values with those observed for pipes of equal diameter but of polished interior surface, he found considerable differences that were not expected. For example, for a cast iron pipe coated with bitumen, a new cast iron pipe and that same pipe with deposits, the resistance coefficient varies approximately as 1, to 1.5, and to 3. It should be remarked that even if the extent of deposit is very small and that the original conduit diameter is essentially maintained, this effect is significant. Accordingly, flow calculations in practice should account for that deposit in all cases. Cast-iron conduits coated with bitumen result in discharges much larger than according to Prony, provided all other parameters remain invariant, whereas used cast-iron pipes lead to much lower discharges. Mr. Darcy also determined by experimentation that Prony's formula involved no large influence on the pipe diameter, because small diameters gave proportionally too large velocities whereas the velocity was too small for large conduit diameters.

Until the great works of our lost colleague, Dubuat was considered the prince of the hydraulicians. He had concluded from a poor experiment the principle that resistance against liquid movement caused by the pipe walls was independent of the pressure exerted on that wall. Mr. Darcy proved that principle without doubt by using very large and greatly different hydraulic heads between 18 and 41 m. The effect of the pipe diameter on the quadratic velocity coefficient is represented by two terms, one varying inversely with the diameter. Mr. Darcy has computed these coefficients and set up a large table at the end of his book where the required sections, the hydraulic heads per 100 meters, and the discharges may be determined for all diameters from the centimeter to 1 meter, and for velocities from 10 centimeters to 3 meters per second. It is important to state that these tables account for new cast iron pipes, implying in practice to divide those discharges by 2. Mr. Darcy took great care to inform the reader of that in the note that precedes the table.

Another table of the book gives all the diameters up to 1 m to allow for immediate prediction of other elements. Accordingly, the slope required can be computed for example when the diameter is given. Mr. Darcy only stated that the hydraulic slope should be doubled as compared to the calculation, or the velocity should be taken by as one-half because of the deposits in used pipes. He also recommended diameters large enough to always account for these deposits, especially for small pipes. What had confused a hydraulician as important as Prony was an experiment involving a pipe of forty-eight centimeters that had been in service for years as compared with observations made with new small cast-iron pipes. Only Mr. Darcy demonstrated the effect of deposits, provided the conduits were considered with their proper conditions of wall surface.

Mr. Darcy not only aimed to solve questions of practical hydraulics but also turned to scientific problems of pipe flow. Accordingly, the velocity across a pipe section was found to decrease from the center to the minimum at the pipe circumference. Mr. Darcy expressed the laws obtained by experience with a formula relating the center velocity to the velocity at a distance R away from the center, and determined the wall velocity as a function of the average velocity. The latter is located at about two thirds of the pipe radius. He also related the maximum velocity to the velocity at distance R away from the pipe axis, and plotted the curve between these two parameters.

A highly unexpected result was that the velocities on two rings of equal distance are inversely proportional to the radii of two pipes of different diameters, but with equal maximum velocity. The hypothesis involving gradual and regular flow was not experimentally confirmed; Mr. Darcy together with Mr. Baumgarten had observed periodic velocity variations, which were also experienced by the height of water jets and cause oscillations in the water level of manometers fixed on pipes. Mr. Darcy considered these interesting observations on pipe flows as a mystery that has so far not yet been revealed. In June 1854, the beautiful memoir of Mr. Darcy containing many important discoveries was the object of a favorable report by Poncelet, Combes and Morin, at Académie des Sciences. The conclusion adopted by that learned society was to insert the paper in the *Recueil des Mémoires des Savants étrangers*, yet it was printed and published only in 1857. That double date explains why I had to start in that review with pipe research before talking on the large work on the *Fontaines Publics de Dijon*, published in 1856.

[Fontaines Publics de Dijon]. I already said that all the problems to solve, all formulas for computation, all precautions to respect, all tunnels to establish, all arrangements to foresee for the distribution of water in the city may be found in that book. Mr. Darcy also introduced the solution of various questions for natural and artificial fountains, such as water filtration, of which most are novel or at least considered under an original point of view, and make up the work on the *Fontaines Publics de Dijon*, a complete treatise on the subject that should take a particular place in the library of all engineers and architects occupied with hydraulics.

Sirs, before examining the interesting part of Mr. Darcy's works, it seems appropriate to remember a great work on public hygiene, namely the improvement of the *Suzon* sewer in Dijon where the ancient cesspool ran along the *Suzon* from north to south. Mr. Darcy had channelized its bed in 1847. The entrance was at Trémouille tower that was later removed close to Cours-Fleury, then it passed below Place Suzon, traversed Musette road, cut Condé road at its entrance to the Forges and Piron roads, and with a very sinusoidal course, it finally discharged into the Ouche by passing below the railroad. The total length of the sewer is 1,328 m, and the slope totals 7.131 m. The sewer entrance is conveyed below the outer foundation of Trémouille tower and blocked with two gates set in 1841, to discharge water from the Suzon for cleaning.

It is true, the sewer covered earlier arches established with different dimensions, shapes and directions, by the riparian owners or by the city under the streets and places, but several parts had remained uncovered. The sewer was nearly everywhere destitute of sills, and it was decided to regularize the width and the slope, and to prevent the deposal of solids from domestic, shop and factory wastes, which were the most pestilential material throughout the city. This last issue concerned the municipal authority, which by a decree of June 11, 1842, approved it and it was subsequently accepted by the administration. The task of Mr. Darcy, the ardent promoter of this important city improvement, was more laborious. All existing arches built more or less to required dimensions were repaired, strengthened and connected as best possible to form a continuous and regular aqueduct. Three new arches were constructed on parts of the cesspool that had remained in the open. Finally its bed was examined, cleaned, and covered in masonry.

In total, the main sewer receives sixteen smaller sewers, and five distribution [overflow] pipes of the Rosoir spring. In addition, Mr. Darcy installed a 16-centimeter pipe with a valve close to the Trémouille tower to clean the aqueduct with a rapid stream of water to avoid infections, especially during the summer period. All these complicated, extended and useful works were finished by 1847. You may realize that the basic idea has since then also been adopted in Paris. Involuntarily, these ideas may even make one think of the antique sewers of Rome that still testify today the power and the foresight of the Romans.

It is now possible to return to the paper published in 1846 [sic, 1856] by our wise colleague that is of such importance both in theory and practice. Among the many novel questions that were solved, we can only mention the principal ideas, in particular those relating to artesian springs. That particular low cost scheme was known for a long time in the *Artois* province. Yet it seems that knowledge was available much earlier in China and in the Orient, particularly in Arabia and along the boundaries of the deserts. Where does the water originate that perpetually feeds the springs so created? Modern geological science answers the question. A probe advanced to the surface of subterranean water under pressure makes water rise and discharge at the surface of the earth like from a hidden reservoir connected with water from surrounding mountains. Evidently, water thus rises up to the level of the reservoir head under these conditions, minus the pressure loss due to the water flow.

By adopting this first principle, Mr. Darcy observed its correctness only if the underground produced a small velocity and the discharge was so small that no appreciable water movement occurred. Without that effect, the friction height had to be added for predicting the height of the jet. This observation permitted Mr. Darcy to solve a problem previously thought to be inaccessible. For a given artesian well, determine whether the feeding stream is infinitely larger than the discharge supplied, or whether the two are nearly equal. To answer, a tube in which the subterranean water may rise is used and its static level is noted. Then that tube is cut a certain length and the quantity of water discharged is noted again. By repetition of that procedure at least four times, a relation between the tube length and the water discharged is established. Using simple hydraulics, the respective hydraulic heads can be determined under which the various flows were operated. If the height for all flow conditions is identical, then the subterranean aquifer is much extended, whereas the opposite conclusion can be drawn otherwise. This theory of artesian springs had found complete confirmation with experiments involving wells drilled at the city of Tour and its vicinities, which establish the dependence where the flows of these wells can vary as mentioned above, by lowering the point of emergence.

These considerations have permitted Mr. Darcy to contest the results the German driller Mr. Kind made for the Passy spring with a 60 centimeters pipe. The reasons for the accident that occurred during the execution of works have not yet finally been solved, however.

For a city water supply there is another, even more important question than just examined, relating to the filtration of raw waters originating from polluted rivers or brooks. Often as in Paris, the essential work is made by a specialized enterprise. If possible it is preferable, to release the population from the duties of water filtration. Mr. Darcy considered that problem also, after many wise engineers had proposed more or less effective means to filter large quantities of water in France, England and elsewhere. In contrast to what resulted from these investigations, he was able to demonstrate experimentally that the quantity of water filtered in a certain time was proportional to the hydraulic head applied to the filter and inversely proportional to the filter thickness, the surface of the filter remaining the same. This principle has widespread application in practice because the filter surface can easily be varied for a certain hydraulic head. Accordingly, a filter basin of 14 m in diameter under a pressure head of 7.5 m renders 15,000 cubic meters of filtered water per day, sufficient for a city of 100,000 souls.

By the way, these filtration vats have offered to the inventive spirit of the man to whom nothing escapes in his research, a simple, cheap and fast method to wash the deposits. When introducing water tangentially along the walls of the vat a circulation in the water flow is created that flushes the previously suspended matter towards the bottom, without harm to the sand in the vat constituting the filter. To discharge the dirty water of normally less than 50 centimeters height after 2.5 hours of filtration, a drain can be opened, and closed again after the outflow is clear. An improvement results when applying the cleaning current longer and directed from the lower to the upper vat direction. These are the ingenious means proposed by Mr. Darcy to render filter operations both efficient and fast, and I thought, sirs, that they had too much importance for not having taken notice.

Here end, besides what I would have liked to tell you about the scientific works of our unfortunate colleague, with all the powers of his genius, with all the resources of science, and with immortal signs that do not only live for us, but which will make him known and assure his glory among the nations, as do his diverse, difficult and useful research. Here appears the complete man, coming after so many others to glean in the field of the hydraulics, because everything on water flow in conduits has been redone such that nothing remains for others in the future.

[Recherches Hydrauliques, 1865]. Unfortunately, it is not known how the work initiated by him on open channel flow will end. The experiments conducted between 1856 and 1857, using the funds sponsored by the ministry of public works along the Canal de Bourgogne at Dijon, led to positive results. These were put together with much care by Mr. Bazin, a young and wise engineer assisting Mr. Darcy in the unfinished part of these works, such that this note will form the most authentic title of priority for our colleague in discoveries so fatally interrupted. [See Darcy and Bazin (1865).]

These experiments have been done in a 580 m long channel along the Canal de Bourgogne. It was divided into two portions. Along the first 130 m, experiments relating the velocity distribution to wall roughness were conducted, with discharges between 100 and 1,200 liters per second. The lower portion of 450 m remained free during theses works, and served for experiments relating to backwater curves, as I will discuss below.

Walls of smooth cement, of longitudinal planks, of stones positioned flat on mortar, of small and larger gravel immersed in a bed of cement were successively used to vary the wall roughness over a wide discharge range. The formulas established using novel coefficients as determined with these experiments differ considerably from those of Etheilwein [sic, Eytelwein] and Prony. Observations in earth and grass covered channels, as often encountered in nature, remain to be made. During the last year, Mr. Darcy had primarily directed experiments on the influence of channel cross-section shape to determine the optimum water flow. He found that the semi-circular section was definitely the best, followed by the trapezoidal and the triangular sections and finally the rectangular channel shape. To have an idea of the significance of that new research, it is sufficient to state that the resistance coefficient for a semi-circular smooth section is only about one-third of that given by Prony in his general formula.

Velocities were also measured at many points of various sections, using the *tube jaugeur*, an improved Pitot tube designed by Mr. Darcy. Observed equal velocity contour lines were regularly distributed over the cross-section, especially for the semi-circular channel shape, where they plot as concentric circles with velocity decreasing from the center towards the wall. The ratio between the maximum and the mean velocity varies between 0.6 and 0.9, and tends naturally more to 1 as the wall roughness decreases. Remarkable experiments were also made in a small channel at Grosbois with a 1:10 bottom slope. For velocities up to 10 m per second, the coefficient of resistance remained almost equal as for normal channel flow.

An important observation relates to the backwater upstream of dams inserted in a channel. For a surface slope of 2 mm per m and for discharges between 100 and 1,200 liters per second, the backwater effect extended over 150 m with an absolutely horizontal surface and an almost vertical jump of 15 cm height. By plotting with great care the surface profiles of a flow in a horizontal channel over 180 m, the free surface slope disagreed with the slope of the gradually varied flow equation as presented in standard hydraulics. That equation must thus be subject to modification.

Finally, the phenomenon of *mascaret* [surge wave] was also investigated. It occurs at river estuaries such as the Seine and Dordogne rivers and corresponds to a great wave overtopping the flow by a height of 1 to 2 m and propagates against the flow upstream, yet is never followed by another wave. By using variable water levels and a number of channels, Mr. Darcy investigated various circumstances of that phenomenon. He finally investigated the case occurring on the Seine and determined its advance at 8 m per second. Accordingly, a complicated phenomenon having been the subject of much speculation, had received a satisfactory and complete answer.

This sirs is the interesting and useful research conducted by our scientist and friend, when death came and made an end to all his conceptions. The great Newton has said of the famous geometer Côtes, who was also taken off at premature age from his laborious analysis: "If Côtes had lived we would know something." We can state without less conviction: "If Darcy had lived, we would have known hydraulics, and this without any reservations." He would not have left any obscure corner in which the flame of experience and formulas would not have penetrated; within a few years it would have been realized. In the same time, the *Académie des Sciences* [of Paris] would have opened their door, because Darcy almost touched its sill. Yet that double glory was refused, and our most sincere expectations remain thus disappointed. God had other orders!

[Illness]. With a deep sorrow, we remember that most of the works of our friend were executed during the reign of a nervous illness of the poorest kind. What were the origins of that cruel affection? Without any doubt, they were worsened by the excessive office work of Mr. Darcy during 1852 and later years. However, the members of his family did not hesitate to bring up the story of a terrible accident as both a physical and moral shock. That was in 1837, during an official reception for the small railroad of Epinac, at the mine of that name. The experimental car descending the slope contained Mr. Chaper then the Prefect, and various administrators and engineers, in which Mr. Darcy naturally had his place. Suddenly he realized that the velocity of the car increased visibly instead of being constant. An interrogation with low voices of the engineer that directed the car left no doubt of the terrible truth: The brake did not work anymore! Soon, a curve appeared along an elevated embankment and it was simple for an engineer to foresee that the velocity was such that the car would inevitably derailed, and after a terrible fall, almost certain death would result. During some moments of fear impossible to describe, Mr. Darcy concentrated on his thoughts, for those who sat around him and for himself – an imminent catastrophe. Fortunately, another neglect providentially saved everything. A forgotten car stood obliquely on the rails. The experimental car hit it under great velocity, pushed violently on the leading corner, overturned and threw all the passengers rudely onto the rails; each lying there more or less squeezed or blessed. Mr. Darcy, for his part, had to stay in bed for one month. Whatever be the influence of that incident in the life of a man, I thought not to omit it to you.

During his last two years, the health of our colleague considerably improved. His family cared for him and rendered the sweet kindness that can be expected only from tenderness. In addition, he followed a hygienic regime given by best councils. In a family troubled over such a long period, there was revival of the highest hopes. One saw a dawn of joy of complete health, when a sudden accidental illness destroyed everything within only a few days, a few hours. Gentlemen what more can you say, to relate to such sadness and to lament in your powers, than to insert the note in your annals and deposit a palm on his tomb?

Appendices: Excerpt of the report presented by Mr. Dumas, mayor of Dijon, to the municipal council, who decided on May 4, 1846 on the means to testify the gratitude of the city towards Mr. Darcy, following the erection of the public fountains

The Mayor, after having recalled the difficult works undertaken by Mr. Darcy, his energetic will and the perseverance with which he had surmounted all the obstacles, added: "This immense task now accomplished, the duty of manifesting the thanks remains to us as the representatives of the citizens. The feelings of generosity that honor our colleague were too much known to us such that we did not dare to offer a financial fee to him, or even the simple fees for his considerable expenses due to the many trips to Paris and the daily surveillance of the works along the aqueduct. A commission of the city found it adequate to present to him by a unique exception and without recurrence in any circumstances, firstly a supply of water that he had procured us so favorably for his house for life, and secondly to stamp a large format medal to commemorate the event that is so memorable for us in establishing the fountains. It will be used by future generations to remember his name as an example of noble devotion. The medal will have the following inscription:

To H. P. G. Darcy
Chief engineer of the Côte d'Or Department
He conceived the project,
Made all the studies,
Pursed until the end of the execution,
Works to which Dijon owes the creation of abundance of its fountains
He did not want to accept financial compensations,
or even fees for his expenses:
The Municipal Council offers a sign of public recognition
To H. P.G. DARCY,
Doubly benefactor of his native city,
by his talents and by his unselfishness

The municipal council adopted the Mayor's proposal:

Article 1. As a testimony of the zeal, the talent and the noble unselfishness of Mr. Darcy, member of the municipal council of Dijon, the municipal council offers him in the name of the citizens a gold medal to commemorate the souvenir. The mayor and a special commission especially created for the purpose will make the presentation.

Article 6. The city will provide Mr. Darcy during all his life to his house free water from the public fountains to satisfy all the needs of his family and his household.

The Mayor of Dijon has ordered on the day of Mr. Darcy's funeral that the Place du Châteaux d'Eaux [Water House Park] will take his name, which was confirmed by the municipal council on the meeting of January 8, 1858 as,

"The council following the public opinion and the righteous steps of Mr. Mayor who has taken the initiative to acknowledge the memory of Mr. Darcy, whose eminent services towards his hometown are attested by great works,"
-Adopts the withdrawal of the sum to be taken from the budget of unforeseen expenses necessary for a plaque with the name of Darcy, to be mounted at Place du Châteaux d'Eaux.
-Declares that the expenses for the two-unit tomb required by the family for the burial of Mr. Darcy will be presented by the city.
-Wishes that a special homage be given to the memory of Mr. Darcy, to be paid by the city,
-And for the study to determine the best realization of this idea, names a commission composed of Messrs. Gaulin, Toussaint, Clerget-Vaucoulcur, Chanoine and Liégeard.

It also wishes that Mr. Mayor forwards a copy of the foregoing deliberation to the family of Mr. Henri Darcy, as a testimony of sympathies and regrets of the municipal council."

Copy of a letter addressed to Mr. Hughes Darcy by Mr. Vernier, Mayor of Dijon and member of the legal corps, dated March 16, 1858.

"I hurry to respond to the question that you addressed on the decision that was taken on January 8 last, relating to the general terms for a special memorial to the memory of your so regrettable and so much regretted brother, by the city council. I am in the frame of mind that convinced the municipal council on the feelings dominating the erection of a marble or a bronze bust, following earlier investigations that proposed to mount it either on or in a public establishment.

I sincerely regret not having been able to be more precise on that favor, of which the city would be happy to find a public notice under the auspices of the General Council of Ponts et Chaussées. Certain differences on details in the commission urge me to postpone a final decision until my return to Dijon, and I wish that this question be one of the few to consider, soon after my return. Needless to add that no other affair of my administration is more at my heart."

Extract of the minutes of the Dijon municipal council.
June 1st, 1858.
Present: Mr. Vernier, President and Mayor, 13 Members and the Secretary.
Mr. Liégeard, Member of the Commission named on January 8 to investigate the best mode to realize the wishes as expressed by the municipal council for a special memorial sponsored by the city to the memory of Mr. Henri Darcy, proposes to place a bust in bronze of that eminent engineer on the top of the monument that was elevated on the entrance of Porte Guillaume, to remember the establishment of the public fountains of Dijon. The correspondent presented also the idea advanced by the commission to install a marble bust of Mr. Henri Darcy with an inscription that reminds to the gratitude of the city.
The Council, by following the correspondent's proposal, decides that the bronze bust of Mr. Henri Darcy, former inspector general of Ponts et Chaussées, be placed on the top of the monument located at Porte Guillaume in order that the

public keeps his memory, and that its execution be confined to Mr. Jouffroy. Mr. Mayor, to this effect, adds the sum of 3,000 francs to the budget of the year 1858. For correct writing, for the absent Mayor, Lejéas.

Translator's Concluding Remarks

Henry Darcy was a benefactor of mankind with his water supply project, engineer able to present designs for the railways taking the Dijon route, improver of the Sologne swamps, introducer of Macadamization in France for modern road construction, and the scientist that investigated pipe flow, improved the Pitot tube and initiated open channel hydraulics research at the Canal de Bourgogne. That three-fold character of a person is considered sufficiently outstanding and motivated the celebration of his 200th birthday both in France and the USA. The present paper is essentially a translation of an obituary by Darcy's friend Caudemberg. He not only give the facts of the eminent engineer but also draws a portrait of a man of his time that was fortunate, but who passed away much too early because of a terrible disease. The story on the unselfish Darcy that refused all financial reward might fascinate not only the hydraulic community but also all those interested in the history of the 19th century.

References

Darcy, H., and Bazin, H. (1865). "Recherches hydrauliques". *Mémoires présentés par divers savants à l'Académie des Sciences de l'Institut Impérial de France* 19:1-652.

Caudemberg, G. de (1858). "Notice sur M. Henri Darcy". *Mémoires de l'Académie Impériales des Sciences, Arts et Belles-lettres de Dijon*, Série 2, 4:109-143.

Charié-Marsaines (1858). "Notice nécrologique sur M. Darcy, Inspecteur générale des Ponts et Chaussées". *Annales des Ponts et Chaussées* 28(1): 90-109.

Petit, H.-A. (1987). "La médaille du Rosoir: Hommage de Dijon à Henry Darcy". *Annales de Bourgogne* 59: 71-80.

APPENDIX: History Of The Darcy Medal

[The following is an abridged translation of a paper by Petit (1987) that details the events in 1846 that led up to the striking of a medal in honor of Darcy.]

On September 6, 1840, a banquet was held in the woods of Messigny, where Victor Dumay then Mayor of Dijon, the Municipal Council, the Prefect, the Lieutenant General, the First President of the Royal Court, and the civil and military authorities were assembled. They inaugurated the 12,695 m long aqueduct that traversed five communities and five hundred sixty proprieties to reach the city of Dijon with fresh and hygienic waters from the Rosoir spring. Gifts were distributed to the workers. At noon, the waters were introduced into the aqueduct and they arrived 3 hours 33 minutes later at the vast reservoir of Porte Guillaume, where citizens were assembled.

The journalists of the day do not seem have been impressed by these works. The newspaper *Courier de la Côte d'Or* on September 8th published just ten lines on the inauguration and dedicated a double column to a morning concert. In the following years, a second reservoir was set up at Montmuzard, 121 hydrants for public and private use were installed and a large fountain containing a 13 m jet was completed at Porte Saint-Pierre [today Place Wilson].

Although the municipal council had ruled that no water should be distributed free to individual households, a deliberation dated May 4, 1846 made an exception for Darcy and his family. In a letter composed by Victor Dumay, two councilors proposed the erection a monument as the just reward for the 'new hero', with these verses added:

> "Darcy, new hero, to your great success,
> The healthier public ends its suffering,
> A water to drink, finally, facing joy,
> Will by that monument given forever."

Once the water supply project was completed, Dumay proposed to the municipal council that the city would have to express its gratitude towards Darcy by appropriate means. A commission of six councilors chaired by Dumay was formed, demonstrating his pride and deep affection. It is not known how that commission worked but it was finally decided to stamp a medal. On March 26 and May 19, 1843, the Mayor addressed two long letters to Comte Sussy, then conservator of the money museum at Paris. In the first letter, the merits of Darcy and the decision to stamp a medal were expressed. Sussy was asked, 1) to propose artists, 2) what procedure to follow, 3) the approximate cost for medal engravings and, 4) the final cost for gold, silver and bronze coins. The medal obverse was specified to contain a structure of the water supply system or an allegory with the city of Dijon. On the reverse, an inscription would refer to Darcy, his works and the time needed to achieve them. In the second letter, Dumay gave a long historical account of the Dijon water supply. He then presented four medal designs and proposed two different inscriptions for the reverse.

In June 1843, the committee made a decision: Darcy should be offered a gold medal by the city, two silver medals should be given to the family, two others should be deposited at the municipal library and the museum, and 300 bronze medals should be distributed. All these medals of 68 mm of diameter would show a view of Rosoir spring and the city of Dijon on the obverse, whereas the reverse should relate to Darcy stating his unselfishness towards his citizens. The gold medal would have a personal indication of votes of thank towards Darcy. A credit of 6,000 francs was put aside for that purpose.

Dumay then contacted Augustin Armand Caqué, medal engraver and official engraver for Napoleon III. He had produced much, but was known more as an excellent technician than a great artist. In a letter dated of March 7, 1844, Dumay asked Caqué to present a design for the obverse. On May 22, 1844, Dumay accepted a fee of 3,000 francs requested by the engraver, but the money would only be paid once an acceptable artistic project was available. On May 5, 1845 the municipal council was assembled and Dumay proposed to ask the secretary of Académie de Dijon for the exact medal inscriptions. These were finally transmitted on September 20, 1845 to the engraver. At the end of the year

Dumay asked authorization for stamping the medal and the work was ordered on January 13, 1846: One gold medal, two silver medals, ten bronze medals and 300 bronze medals with slightly different inscriptions. In total 315 medals were ordered, as decided by the municipal council on May 4, 1846. The gold medal has a weight of 242.8 gm, and its price was 860.49 francs of which 10.20 francs were paid for the stamp. A leather case with four small holes cost an additional 15 francs, and was presented in a wooden box priced 1 franc each.

On May 4, 1846 there was a great assembly of Dijon municipal council. After a long speech by Dumay, the following was decided:

1. In the name of the citizens, a gold medal would be presented to Henry Philibert Gaspard Darcy, to acknowledge his zeal, his talents and his noble unselfishness. The Mayor accompanied by a delegation would make the presentation. Two identical silver medals would be presented to Mrs. Darcy, mother of the engineer, and to Hughes Iéna Darcy, Prefect of the Gard department, Darcy's brother. A list followed with the names of the recipients of the bronze medals.
2. A delegation including the Mayor, two assistants and six councilors went shortly after 4 p.m. on May 5, 1846 to Darcy's house. The event was hardly acknowledged in the local press.
3. In a meeting on May 22, 1846, Darcy, himself a municipal councilor, expressed thanks towards the council. Dumay responded that it was only an act of justice.
4. Some days earlier, the 300 bronze medals had arrived for which the expenses were 528 francs. Dumay asked from the Prefect an authorization to add 72 francs from an emergency budget since the original 6,000 francs was exhausted.
5. A list of 280 names was made to which the bronze medals were sent. The remaining medals were given out by hand.

Darcy was unable to avoid the strict directives to place an additional bronze medal in the foundation of the Montmuzard reservoir. On May 25, 1846 Dumay wrote him, "Place it into an envelope of cloth or wool and cover all sides with charcoal in a circular cavity at least 1 decimeter depth, vertically along the stone posed in the superior portion, in a south-western direction." Dumay stated that all these measures were taken to keep the souvenir intact.

Figure 3 presents the medal. The obverse shows scenes from Dijon with the current Place Darcy fountain in the center. The inscription reads, "The Rosoir spring is brought to Dijon". The reverse is all text and reads,

> "The municipal council of Dijon, by act of March 5, 1834, realizing that projects rested in vain for three centuries to supply hygienic waters to the city, in the legacy of Abbé Audra, contributed to the expenditure for the construction of the circular reservoir containing 22,000 hl and the subterranean aqueduct of 12,965 m length and discharging 8,000 liters per minute, that were initiated on March 21, 1839 and completed on September 6, 1840, after the design and under the able direction, and the yet unselfish H. P. G. Darcy, Chief Engineer of Côte d'Or department.

The last letter concerning the medal was dated February 15, 1847. Thus more than five years had passed. In contrast, the reservoir on Place Darcy was completed within 128 days.

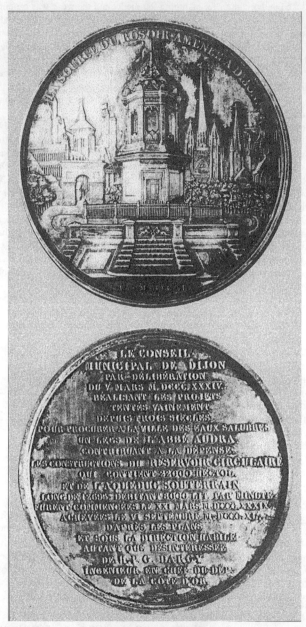

Figure 3. The Darcy Medal.

The Place of Darcy's Law in the Framework of Non-Equilibrium Thermodynamics

P.H. Groenevelt[1]

Abstract

Darcy's Law (1856) was the fourth of the famous nineteenth-century "Laws" of transport. It followed Fourier's Law (1822) for heat transport, Ohm's Law (1827) for the transport of electric charge and Fick's Law (1855) for the transport of dissolved chemicals. All these linear, homogeneous "laws" are energy dissipation equations, relating driving forces to their (gut-feeling) conjugated fluxes. In a natural porous body, such as a soil, all these transport processes usually occur simultaneously. Starting with the Gibbs (1876) equation, one can develop an expression for the total dissipation of energy in such bodies. Products of the driving forces and their conjugated fluxes appear as separate "blobs" of energy dissipation. It was Onsager (1931), who postulated that if one sets every flux appearing in the dissipation equation as a linear homogeneous function of each of the driving forces appearing in that same equation, one finds that the matrix of phenomenological coefficients is symmetrical. The equality of the cross-diagonal coefficients is known as Onsager's Reciprocal Relations. For this work Onsager received the Nobel Prize for chemistry in 1968. Within this whole framework of transport equations, Darcy's Law plays a central role, and, in the presence of driving forces other than the gradient of the hydraulic head, it is augmented by terms accounting for (capillary) osmosis, electro-osmosis and thermo-osmosis. Simultaneously all other fluxes now contain a term for which the driving force is the gradient of the hydraulic head: reverse osmosis (salt sieving), streaming current and heat filtration, respectively. The Philip and De Vries (1957) theory of simultaneous heat and water transport in soils is not founded in thermodynamics, and the Onsager Reciprocal Relations are lost. The misunderstanding of coupled transport has lead some researchers to erroneously conclude that their observations are "non-Darcian". It can be shown that a linear coupled transport process, when "built" into the linear Darcy equation, turns the latter non-linear (Bolt and Groenevelt, 1969).

It is worth noting that this special year, in which we celebrate the 200th anniversary of Darcy's birth, also marks the 100th anniversary of the death of Josiah Willard Gibbs (1839-1903) and the birth of Lars Onsager (1903-1987).

---

[1] Professor, Department of Land Resource Science, University of Guelph, Guelph, Ontario, Canada, N1G 2W1. Phone: 519-824-4120 X53585

Introduction

At the time of the birth of Henry Gaspard Philibert Darcy on June 10, 1803, the branch of science known as Mechanics, had already risen to great heights, largely due to the works of Isaac Newton. The branch of science called Thermodynamics, however, was nowhere in sight. When Darcy was 19 years old, in 1822, another great French scientist, Jean Jacques Batiste Fourier, formulated the first of a series of fundamental transport equations, that would see the light during the next four decades. Fourier's Law for heat transport consists of a linear homogeneous relationship between the temperature gradient, as the driving force, and the heat flux as the response exhibited by a homogeneous medium. By the time that Darcy graduated as engineer from the Ecole Polytechnique in Dijon in 1826 (Philip, 1995), he must have heard of Fourier's famous book "La Theory de la Chaleur" (1822). Thirty years later, when Darcy formulated his now famous homogeneous linear law of water transport through a porous medium, the principle of this type of equations was well known. In the meantime, Ohm (1827) and Fick (1855) had formulated their homogeneous, linear transport equations for electric charge in a solid and a solute in a liquid.

Basis of Darcy's Law

The foundation of Darcy's Law lies in the expression for the viscous energy dissipation in a moving fluid. On a molecular scale the Navier-Stokes equation formulates the force balance from which the dissipation function can be derived,

$$\Phi = T\sigma = -\tau : \nabla v \qquad (1)$$

where Φ is the amount of energy dissipated per unit time per unit volume, T is the absolute temperature, σ is the entropy production, τ is the viscous stress tensor, ∇v is the gradient of the velocity vector and the colon is the tensorial operator, that contracts the two tensors, τ and ∇v, into a scalar.

This dissipation function is formulated for a small pocket of water molecules (the "Navier-Stokes scale"). When the boundaries of the system, in which the fluid is flowing, are known precisely, integration of the Navier-Stokes equation leads to a macroscopic, linear, homogeneous transport equation, relating the gradient of the fluid pressure to the fluid flux. Poiseuille's "Law", for a pipe, is a prime example of a derived dissipation equation. For a porous body, such as a soil, Darcy's "Law" is the equivalent of Poiseuille's Law. It is only because of the intricate, impossible to describe, boundary conditions, that Darcy's Law cannot be "derived". Therefore, Darcy's law is, what is called in the discipline of Thermodynamics, a "phenomenological equation", describing the dissipation of energy, such that the product of the flux and the force is equal to the energy dissipation. The energy dissipation is equal to the product of the entropy production and the absolute temperature of the system.

The other 19th century phenomenological equations follow the same recipe, except for one, Fick's "Law" for the diffusion of a dissolved chemical. Fick chose as the driving force the gradient of the solute concentration. The product of the negative gradient of the solute concentration and the diffusion flux is not equal to the energy dissipation associated with the mixing of the solute in the fluid. The driving force should be the negative of the gradient of the chemical potential or the osmotic pressure. Fick is not to blame for this, because the chemical potential was not heard of until Josiah Willard Gibbs "invented" it in 1876 and the osmotic pressure could not be measured until Jacobus van 't Hoff showed how to do it in the late1890's. Thus, in the framework of non-equilibrium thermodynamics (NET), the driving forces must be gradients of potentials (energy levels) and *not* concentrations. It is precisely for this reason that the mathematical development in soil physics during the 1950's, spearheaded by J. R. Philip (1955), introduced the opportunity for errors. In order to obtain a diffusion-type second order partial differential equation for the movement of water in soil (a sort of Fick's second law" equation), the driving force, as proposed by Darcy, was modified into a gradient of the volumetric water content. Thermodynamically that is a dangerous move. As long as Philip proceeded very carefully, exploiting the error function solutions, that were available for Fick's second law, the results are in order. In extending the development to coupled processes, however, the footing in thermodynamics is lost. Thus, in the theory of simultaneous heat and water transport as proposed by Philip and De Vries (1957), the immediate consequence is that the Onsager Reciprocal Relations (ORR) are lost. ORR means that the coupling coefficients are equal.

Raats (1975) presents a detailed analysis of both the Philip and De Vries theory and the thermodynamic theory as developed by Groenevelt and Kay (1974). He comes to the conclusion that the Philip and De Vries theory simply does not fit in the framework of non-equilibrium thermodynamics. In his most recent article in the "The Philip Volume", Raats et al. (2002) state: "Using this theorem (i.e. Meixner's), it can be shown in particular that if one follows Philip and De Vries and chooses the gradients of water content and temperature as the driving forces, the Onsager relations are no longer satisfied. But this is no cause for alarm: the expressions for the fluxes, with the gradients of the water content and temperature as the driving forces as formulated by Philip and De Vries, remain valid just the same. These expressions merely do not fit in the framework of thermodynamics of irreversible processes." All transport processes, whether they are conjugated or coupled, are "irreversible" processes. If the expressions (i.e. transport equations) do not fit in the framework of non-equilibrium thermodynamics, one finds that the sum of products of fluxes and driving forces does not add up to the total energy dissipation due to all occurring processes. This is a central requirement in NET. The violation of Onsager's reciprocity simply means the loss of the equality of the coupled transport coefficients and therefore the requirement to measure one more coefficient than is necessary. Often, one of these two coupling coefficients is much more difficult to measure than the other and thus the ORR can be extremely useful. A more serious problem arises when the simultaneous transport of heat and water is accompanied by simultaneous transport of solutes and electric charge. This is universally the case in clay soils. The matrix of 16 coefficients (four straight or conjugated and twelve coupled transport coefficients) then becomes rather messy.

The Framework of Linear, Non-Equilibrium Thermodynamics

The Gibbs equation is the foundation of the development of linear non-equilibrium thermodynamics. Each of the terms in the Gibbs equation is differentiated with respect to time and subsequently; each term is replaced by a conservation (continuity) equation. Each conservation equation contains a divergence term and a source/sink term (Groenevelt and Bolt, 1969). The source term in the conservation equation for entropy is the entropy production. Isolating the entropy production and multiplying it by the absolute temperature, one finds the energy dissipation function, Φ.

The complete energy dissipation function should represent all energy dissipating processes (nothing overlooked and nothing counted double). The expression for Φ usually consists of the sum of a number of products of fluxes and forces. Fluxes and forces can be rearrangement as long as to total sum of energy dissipation remains intact. One such expression for Φ reads:

$$\Phi = T\sigma = - j_q \text{ grad } T - j^V \text{ grad } H - j^D \text{ grad } \pi - j^E \text{ grad } E \tag{2}$$

where j_q is the caloric (Fourier) heat flux; j^V is the (Darcy) flux of the soil solution; j^D is the (Fick) diffusion flux and j^E is the electric current (usually indicated by capital I). Grad H is the gradient of the hydraulic head, grad π is the gradient of the osmotic pressure and grad E is the gradient of the electrostatic potential.

Non-equilibrium thermodynamics now postulates that each flux, occurring in the energy dissipation function is a linear, homogeneous (no intercept) function of all forces occurring in the same equation for the total energy dissipation. Thus:

$$j_q = - L_{TT} \text{ grad } T - L_{TV} \text{ grad } H - L_{TD} \text{ grad } \pi - L_{TE} \text{ grad } E \tag{3}$$

$$j^V = - L_{VT} \text{ grad } T - L_{VV} \text{ grad } H - L_{VD} \text{ grad } \pi - L_{VE} \text{ grad } E \tag{4}$$

$$j^D = - L_{DT} \text{ grad } T - L_{DV} \text{ grad } H - L_{DD} \text{ grad } \pi - L_{DE} \text{ grad } E \tag{5}$$

$$j^E = - L_{ET} \text{ grad } T - L_{EV} \text{ grad } H - L_{ED} \text{ grad } \pi - L_{EE} \text{ grad } E \tag{6}$$

where L_{ik} indicates the coefficient of transport of i, resulting from a gradient in k. The diagonal terms of the right hand sides of equations (3) to (6), the ones with the L_{KK} coefficients, represent the "straight", well-known transport processes, the "laws" of Fourier (1822), Darcy (1856), Fick (1855), and Ohm (1827), respectively.

Coupled Transport Processes

All other (off-diagonal) terms represent coupling processes, and their coefficients, L_{ik}, are known as coupling coefficients. They come in pairs of twins, and are located symmetrically across the diagonal. They represent all possible coupling phenomena in of the four driving forces occurring in this scheme.

Almost all these coupled transport processes do occur in soils, particularly in clay soils and clays. In other branches of science many of these coupled processes are well known, such as the Peltier effect (L_{TE}), the Seebeck effect (L_{ET}), the Soret effect (L_{DT}) and the Dufour effect (L_{TD}).

Onsager Reciprocal Relations

Lars Onsager (1903-1976) postulated the equality of the twin couplings coefficients:

$$L_{ki} = L_{ik} \tag{7}$$

For this work he received the Nobel Prize for Chemistry in1968. The foundation of his work lies in the works the great thermodynamicist, Josiah Willard Gibbs (1839-1903). Onsager sat in the Chemistry Chair at Yale University, the same Chair that had been occupied by Gibbs for 30 years.

Darcy-Coupled Processes (all coefficients containing the subscript V)

Thermo-osmosis and Thermofiltration *(L_{VT} and L_{TV})*. Both these phenomena are very common in soils. The transport of heat due to a water potential gradient in the absence of a temperature gradient and the transport of water due to a temperature gradient in the absence of a water potential gradient are occurring constantly in the soil. In unsaturated soil they are due to evaporation and condensation [the selection mechanism here is the heat of vaporization/condensation]. In saturated soils the magnitude of these phenomena is very small [the selection mechanism is now the heat of wetting]. In frozen soils they are caused by freezing and melting. [here the selection mechanism is the heat of freezing/melting and solidification/sublimation]. These phenomena and the values of the two coefficients are discussed in the literature by Kay and Groenevelt (1974).

Capillary Osmosis and Reverse Osmosis or Salt Sieving *(L_{VD} and L_{DV})*. L_{VD} (osmosis, also called capillary osmosis) and L_{DV} (reverse osmosis or salt sieving). These phenomena are very common in soils. The magnitude is directly related to the clay content of the soil. [The selection mechanism lies with the electrical double layer]. Clay particles expel negative ions and therefore, they expel dissociated salts (negative adsorption). The longstanding conflict as to whether the osmotic pressure (potential) should be added to the hydraulic potential of water to produce the "total" potential of water, the gradient of which then serves as the driving force on the water is here resolved. As the value of L_{VD} is almost always smaller then the value of L_{VV} (except for a perfectly semi-permeable membrane), the concept of the "total" potential is useless. The ratio L_{VD} over L_{VV} is known as the reflection coefficient.

Electro-Osmosis and Streaming Potential *(L_{VE} and L_{EV})*. These electrokinetic phenomena also find their cause in the existence of electrical double layers. The most commonly observed result of the effects is the streaming potential in clay soils.

Non-Darcian Behavior

Linear homogeneous equations are consistent with "Linear Non-equilibrium Thermodynamics". In soils, coupling phenomena are common. One of the mechanisms causing coupling is the selectivity caused by the presence of electrical double layers. Clay particles are negatively charged and accompanied by counter ions swimming in the soil solution. Not recognizing the coupling a single straight process may seem non-linear. This has often been reported, particularly for slow processes in clay soils (Hadas, 1964; Kutilek, 1969; Olson, 1965; Swartzendruber, 1963 and 1967).

The linearity of Darcy's law does indeed break down when a fluid is forced through a porous media at great speed, under a high pressure gradient. The non-linearity is then due to an "additional" form of energy dissipation, viz. turbulence. This, for all practical purposes lies outside the field of soil physics. The non-linear form of Darcy's equation is known as the Forchheimer equation. One then has entered the realm of non-linear non-equilibrium thermodynamics. This domain has recently been developed by Prigogine and coworkers (Glansdorff and Prigogine, 1971).

Conclusion

Darcy's law represents a form of energy dissipation, which, in the energy dissipation function, occupies a place similar to those occupied by Fourier's, Ohm's and Fick's laws. "Non-Darcy" flow at low gradients is shown to result not from errors in Darcy's Law, but instead the failure to consider other gradients.

References

Bolt, G. H., and Groenevelt, P. H. (1969). "Coupling Phenomena as a possible cause of "Non-Darcian" Behavior." *Bull. Int. Assoc. of Scientific Hydrology* XIV, 2:17-28.

Darcy, H. G. P. (1856). *Les Fontaines Publiques de la Ville de Dijon.* Dalmont, Paris.

Gibbs, J. W. (1961). *The scientific papers of J. Willard Gibbs.* Vol. I Thermodynamics. Dover, New York, 434 pp.

Glansdorff, P., and Prigogine, I. (1971). *Thermodynamic Theory of Structure, Stability and Fluctuations.* Wiley, London.

Groenevelt, P. H. (1971). "Onsager's Reciprocal Relations." *Search* 2: 264-267.

Groenevelt, P. H., and Bolt, G. H. (1969). "Non-Equilibrium Thermodynamics of the Soil-Water System." *J. of Hydrology* 7:358-388.

Groenevelt, P. H., and Kay, B. D. (1974). "On the interaction of water and heat transport in frozen and unfrozen soils. II The Liquid Phase." *Soil Sci. Soc. Amer. Proc.* 30:400-404.

Hadas, A. (1964). "Deviations from Darcy's law for the flow of water in unsaturated soils." *Israel J. Agri. Res.* 14:159-168.

Kay, B. D., and Groenevelt, P. H. (1974). "On the interaction of water and heat transport in frozen and unfrozen soils. I: Basic Theory; The Vapor Phase." *Soil. Sci. Soc. Amer. Proc.* 38:395-400.

Kutilek, M. (1969). "Non-Darcian flow of water in soils (laminar region)." *IAHR Symposium,* Haifa, Israel.

Olson, H. W. (1965). "Deviations from Darcy's law in saturated clays." *Soil Sci. Soc. Am. Proc.* 29:135-140.

Onsager, L. (1931). *Phys. Rev.* 7:405 and 38:2265.

Perfect, E., Groenevelt, P. H., and Kay, B. D. (1991). "Transport phenomena in frozen porous media." pp. 243-270. *In* : Bear, J. and Corapcioglu, M.Y. (eds.). *Transport Processes in Porous Media.* NATO/ASI. Kluwer Academic Publishers, Dordrecht, The Netherlands.

Philip, J. R. (1955). "Numerical solutions of equations of the diffusion type with diffusivity concentration dependent". *Trans. Faraday Soc.* 51: 885-892.

Philip, J. R. (1995). "Desperately Seeking Darcy in Dijon." *Soil Sci, Soc. Am. J.* 59:319-324.

Philip, J. R. and De Vries, D. A. (1957). "Moisture movement in porous materials under temperature gradients." *Trans. Am. Geophys. Union* 38: 222-232.

Raats, P. A. C., Smiles, D. E and Warrick, A W. (2002). "Contributions to Environmental Mechanics: Introduction." Pp 1-28. In: *Environmental Mechanics. Water, Mass and Energy Transfer in the Biosphere: "The Philip Volume", Geophysical Monograph 129.* American Geophysical Union, Washington, D.C.

Raats, P. A. C. (1975). "Transformations of fluxes and forces describing the simultaneous transport of water and heat in unsaturated porous media." *Water Resources Res.* 1:938-942.

Swartzendruber, D. (1963). "Non-Darcy behavior and the flow of water in unsaturated soils." *Soil Sci. Soc. Am. Proc.* 27:491-495.

Swartzendruber, D. (1967). "Non Darcian movement of soil water." *Int. Soil Water Symposium,* Prague, pp. 207-222.

Darcy's Law from Water to the Petroleum Industry: When and Who?

Paolo Macini[1], Ezio Mesini[1]

Abstract

Henry Darcy, in an appendix of his work "*Les Fontaines Publiques de la Ville de Dijon*", described the law governing the flow of water through a saturated sand filter. In this law, flow velocity is correlated to hydraulic gradient by means of a linear proportionality constant K, defined as "permeability" (*perméabilité*). In using this terminology, Darcy apparently followed the practice of hydraulic engineers of his time. Indeed, the permeability concept formulated by Darcy is known today as hydraulic conductivity, a flow parameter describing both the physical properties of the porous medium and the characteristics of the fluid. Only in the 1950s did M. K. Hubbert provide a complete theoretical foundation to Darcy's empirical expression, deriving it from the general Navier-Stokes equations. It is important to remember that Darcy's law was already utilized in many technical fields, and especially in the petroleum industry. In the petroleum industry Darcy's law is formulated in terms of pressure gradient and generalized for oil and gas flow, which led to the concept of multiphase flow. This was accomplished by separating the properties of the rock from that of the fluid by manipulating the proportionality constant K, thereby obtaining a "generalized" law. By doing so, permeability becomes a property of the porous media, dictated by pore geometry alone. The origin of the generalization of Darcy's law can be traced back to the work of American geoscientists at the end of the 1920s. The present paper proposes to establish when and by whom Darcy's law was first generalized and made suitable for petroleum reservoir engineering. In particular, the paper retraces some historical developments of the permeability concept by reviewing the earliest studies that contributed to the popularization of the generalized form of Darcy's law for fluid flow other than water.

Introduction

Intellectual curiosity, enthusiasm and ignorance were the driving forces that led the authors to write this paper. Frequently in academic courses on hydrogeology and groundwater hydrology, the common form of Darcy's law is given as,

---

[1] University of Bologna, Dept. of Chemical, Mining & Environmental Engineering, Via Risorgimento 2, Bologna 40136 Italy. paolo.macini@mail.ing.unibo.it; ezio.mesini@mail.ing.unibo.it

or, in differential form,

$$Q = KA\frac{\Delta h}{L} \quad (1a)$$

$$q = -K\frac{dh}{dL} \quad (1b)$$

where Q is volumetric flow rate [m^{3}s^{-1}], K is hydraulic conductivity [ms^{-1}], A is flow area perpendicular to flow lines [m^2], L is flow path length [m], Δh is change in hydraulic head over path L [m], and q is Darcy flux, *i.e.*, the volumetric flow per unit area [ms^{-1}]. The hydraulic head at a given point is the sum of the pressure head and the elevation: h = (z + p/ρg), where z is elevation measured from a datum [m], p is water pressure [Pa], ρ is water density [kgm^{-3}], and g is acceleration of gravity [ms^{-2}].

In petroleum reservoir engineering, the most common form of Darcy's law is,

or, in vector form,

$$Q = \frac{k}{\mu}A\frac{\Delta(p+\rho gz)}{L} \quad (2a)$$

$$\vec{q} = -\frac{k}{\mu}\text{grad}\Phi \quad (2b)$$

where $\vec{q}$ is velocity (vector) of Darcy flux [ms^{-1}], k is absolute permeability tensor [m^2], μ is fluid viscosity [Pas] and $\Phi = p + \rho gz$ is fluid potential [Pa]. Equation 2 is also known as the generalized Darcy's law, which can be applied when the saturating fluid is different than water at standard conditions.

From the above equations, it is simple to derive an expression that links the two forms of Darcy's law, by splitting the hydraulic conductivity in terms of absolute permeability and fluid properties,

$$K = k\frac{\rho g}{\mu} \quad (3)$$

Here several questions arise. When was Darcy's law "generalized" and who formalized this expression? Was the idea of a single researcher or was it the result of a gradual development supported by laboratory experiments? Are there papers that explain the intellectual thought process that led to the generalized expression? In the authors' opinion, this event is important because it underscores and emphasizes a new concept of permeability. Opinion and usage have gradually converged toward an agreement that the term "permeability" shall pertain to a property of the solid only, and so permeability of a porous medium is the same for different fluids. This is a direct consequence of the generalization of Darcy's law that introduced the concept of relative permeability in case of multiphase flow and the extension to unsaturated soils, where the hydraulic conductivity becomes a function of the water content.

Historical background

The Italian scientist Bernardino Ramazzini (1691) was the first to recognize the true nature of the artesian wells of Modena, in the Po Valley of Italy, and he identified the flow of water through sands. A few years later, Antonio Vallisnieri (1715) completed Ramazzini's observations and worked out the cycle of underground fluid flow in modern "hydrogeologic" terms. The earliest mention of flow measurements through porous media occurred in 1856, when Henry Darcy published the results of his classic experiment on water flow through sand. The experiment is described in an appendix of the famous study on the water supply system of the city of Dijon, France (*Les Fontaines Publiques de la Ville de Dijon*, 1856). In the appendix, entitled *Détermination des lois d'écoulement de l'eau à travers le sable* (Determination of the laws of water flow through sand), Darcy describes the experiment that he had carried out to quantify water flow through a saturated sand filter. The results of the experiment were generalized into a "law", empirical in nature, which nowadays bears his name. The law relates flow velocity to hydraulic gradient through a linear proportionality constant that Darcy named *perméabilité*, probably following a common terminology adopted for such a concept by the hydraulic engineers of his time (Darcy stated that "k is a coefficient dependent upon the permeability of the sand layer"). In fact, this proportionality constant contains information on both the nature of the fluid and on the geometry of the porous medium. The proportionality is known today as the hydraulic conductivity, and has the dimension of velocity. Since its discovery, Darcy's law has been found valid for any Newtonian fluid in saturated porous media, and can be adjusted to account for unsaturated and multiphase flow.

Darcy spent his entire professional career at the Corps des Ponts et Chaussées (Corps of Bridges and Roads), a government agency and a fraternity of engineers with a strong sense of brotherhood and an influential status in those years. Among others, Chezy, Pitot, de St. Venant and Navier were members of the Corps (Freeze, 1994). A contemporary colleague of Darcy at the *Corps* was Arsene Juvenal Dupuit, a theoretical oriented civil engineer who dealt with problems of open channel flow and seepage through soils. In his work of 1863, entitled *Etudes théoriques et pratiques sur le mouvement des eaux dans les canaux découverts et à travers les terrains perméables* (Theoretical and practical studies on water flow in open channels and through permeable soils), Dupuit derived an expression for movement of water through soils that proved to be equivalent to Darcy's empirical law (Brown, 2002). Moreover, by integrating the equation of motion over a radial domain, Dupuit derived solutions for steady flow in a confined and phreatic aquifer. Immediately following Darcy's intuitive contribution, research was "actively pursued by engineers in Austria, France and Germany to solve practical problems of groundwater seepage that were of interest to civil engineers during the second half of the 19th century. These engineers restricted themselves to steady-state flow systems. Whereas the transient process involves both (conductance and storage), the steady-state problem involves only the conductivity parameter" (Narashiman, 1998). In the early 1900s Adolph Thiem and his son Gunther of Germany derived the expressions for the steady radial flow of water in confined and unconfined aquifers (Thiem, 1906), probably unaware of the earlier contributions of Darcy and Dupuit.

In the United States, the earliest recorded data relating pressure to fluid flow through porous rocks seem to be those of Newell in 1885 (Plummer & Woodward, 1936), and later the first comprehensive studies were by King (1899) and Slichter (1899), who measured the rate of flow of water and air through consolidated and unconsolidated sands. Slichter, working at the University of Wisconsin, identified the geometric and viscous drag components of hydraulic conductivity. Slichter also analyzed the steady flow of water through natural porous media: he was probably unaware of Forchheimer's work and derived the Laplace equation independently. (The Austrian engineer Phillip Forchheimer was amongst the earliest to recognize the concept of isopotential lines and streamlines in groundwater flow; in 1898 he described the steady flow of groundwater by formally writing the Laplace equation.) From 1900 to 1920 there appears to have been little interest in permeability. Groundwater hydrology was still in its infancy and petroleum engineering was far from being established. Moreover, interactions and exchange of papers between researchers were scarce. The cases of the Thiem's and Slichter's publications are emblematic. In the same years, Terzaghi (1924), who is considered the founder of the discipline of soil mechanics, experimentally studied the deformation of water-saturated clays and established relationships among external stresses, pore fluid pressure and deformation.

Almost all the theoretical and applied studies of fluid flow in porous media were in the domain of groundwater. In 1914, the U.S. Bureau of Mines (USBM) established the Petroleum Division Laboratories in Bartlesville, Oklahoma, which can be viewed as the beginning of reservoir engineering, a branch of petroleum engineering, with its own physiognomy as a new science. However, the American Petroleum Institute Division of Petroleum Engineering was not established until 1927. In fact, in 1925 it was already well known that "From the standpoint of getting oil from the reservoir sands, our present methods of extraction are very inefficient. From all data compiled it is evident that about three times as much oil remains in the sand after economic production has ceased as was originally produced." (Brewster, 1925). The problem was to increase the recovery factor, to find the best geometry of well spacing and to study waterflooding operations. Lester C. Uren's *Petroleum Production Engineering* (1924) was the first textbook on this subject. Prior to this book, only a handbook, J. R. Suman's *Petroleum Production Methods* (1921), was available. At that time there was little clarity on the difference between porosity and permeability, as exemplified by Wyckoff's statement: "porosity and permeability have even been frequently confused" (Wyckoff *et al.*, 1933a).

In the 1920s the increased demand for oil led the USBM to begin a series of experiments to determine the characteristics of oil and water flow in the subsurface. Mills (1921) studied the relationship between the flow of liquids and texture and structure of sands, and Melcher (1920) developed and described a method for finding the density and porosity of oil sands. He also conducted a quantitative study of the effects of pressure changes on rate of flow of crude oil through Oklahoma oil sands (Melcher, 1925). Since 1930, the importance of flow measurements for the production of oil and gas has been emphasized by a large number of investigations. Among these the works of Tickell (1928), Stearns (1928), Fettke & Mayne (1930), Fettke & Copeland (1931), Barb (1930), Barb & Branson (1931), Nutting (1926, 1927, 1929, 1930), Botset (1931), Muskat & Botset (1931), Nevin (1932), Tickell *et al.* (1933), Moore *et al.* (1933), Wyckoff *et al.* (1932, 1933a, 1933b, 1934) have contributed much to the knowledge of the technique of permeability measurements and to the laws of liquid and gas flow through porous media.

In the 1920s to early 1930s, a strong need was felt for the experimental validation of Darcy's law through various porous materials, both natural and artificial with different fluids and gases. Turbulent flow (non-Darcy conditions) was studied as well. As a matter of fact, in most of the research work of this time there was no statement or even mention of Darcy's law, as in the case of the authoritative treatise by Meinzer (1923), where a unit of measure of permeability is also defined (the "meinzer", equivalent to $5.42 \cdot 10^{-10}$ cm^2). Paramount was also the search for possible general relationships between measurable statistical parameters of sands (such as porosity, coarseness, shape, *etc.*) and permeability. It is important to remember that the first standard practice of permeability measurement was proposed in 1933 (Wyckoff *et al.*, 1933-a), and the standard API procedure for permeability determination was issued in 1935 (API, 1935). Soon it was recognized that the validity of Darcy's well-known empirical relationships could be extended to fluids other than water at standard conditions, and that an expression could be developed that properly embodied all of the relevant variables. To this end, it was necessary to account separately for both the influence of the fluid and geometrical properties of the porous medium by splitting the concept of hydraulic conductivity K into its physical components. In doing so the foundations for the concept of absolute permeability k and the study of multiphase flow were laid.

Again, in these years it seemed that research in the field of petroleum reservoir engineering went its own way, characterized by a strong orientation towards practical applications. Contact with the scientific "community" of groundwater hydrologists (if such existed at all) were poor and infrequent. This will be criticized by Hubbert (1969), one of the most important scientists working in both fields who contributed to the theoretical formalization of Darcy's law (Hubbert, 1940, 1956).

When and Who?

Slichter (1899), probably the earliest American scientist to study the flow of fluids in porous media, investigated the flow of a gas-free liquid through spherical-grain sand (Schriever, 1930). By theoretical considerations involving a large number of approximations he derives (rearranged),

$$V = \frac{Bd^2}{C\mu(1-\varphi)} A \frac{P}{L} t \tag{4}$$

where V is volume of liquid delivered in time t, B and C are constants determined from geometrical relations and porosity, d is sand grains diameter, μ is fluid viscosity, φ is and porosity, A is cross sectional area of the sand column, P is pressure drop along the sand column, and L is length of the sand column. Apparently, this is the formula used in all later experimental studies of fluid flow in porous media.

At that time the Poiseuille equation (1842) for viscous flow in circular tubes was well known. It can be written as,

$$v = \frac{d^2 \Delta P}{32\mu L}, \tag{5a}$$

or, in terms of volumetric flow,

$$Q = \frac{\pi r^4 \Delta P}{8\mu L} \quad (5b)$$

where v is fluid velocity [m s^{-1}], d is tube diameter [m], ΔP is pressure loss over length L [Pa], μ is fluid viscosity [Pa s], L is length over which pressure loss is measured [m], and Q is flow rate [m^3 s^{-1}]. If the porous medium is considered to be a bundle of tubes such that the flow could be represented by a summation of the flow from all the tubes, then the total flow would be,

$$Q_t = n \frac{\pi r^4 \Delta P}{8\mu L} \quad (6)$$

where n is the number of tubes of radius r. If the rock consists of a group of tubes of different radii, then Eq. 6 can be rewritten as,

$$Q_t = \sum_{j=1}^{k} n_j \frac{\pi r_j^4 \Delta P}{8\mu L} = \frac{\pi \Delta P}{8\mu L} \sum_{j=1}^{k} n_j r_j^4 \quad (7)$$

where n_j is number of tubes of radius r_j and k is number of groups of tubes of different radii. The above equation reduces to,

$$Q_t = C \frac{\Delta P}{\mu L} \quad (8)$$

with,

$$C = \frac{\pi}{8} \sum_{j=1}^{k} n_j r_j^4 \quad (9)$$

where C is treated as a flow coefficient for the particular group of tubes. If the fluid conducting channels in a porous medium could be defined in terms of the size and number of each radii, then one could use Poiseuille's flow equation for porous media. As there are numerous tubes and radii involved in each segment of porous rock, it is impractical to measure these quantities (Amyx, 1960). However, both equations 4 and 8, when compared to Darcy's empirical relationship (Eq. 1), suggest, at least intuitively, that the flow can be expressed in terms of pressure gradient (at least dimensionally), and that the conductivity term can be untied from fluid viscosity leading to Eq. 3. After Slichter, all authors reviewed in this study did not express this concept clearly in their writings, at least until Wyckoff et al. (1933). Amongst the forerunners, it is important to remember the works of P. G. Nutting (1927, 1930). Nutting, a geophysicist who worked for the United States Geological Survey (USGS), contributed towards the clarification of certain mathematical aspects of viscous fluid flow in pores and fractures. He paved the way for the subsequent formalizations of the permeability concept by Wyckoff et al. (see below). In a 1927 paper, Nutting wrote Poiseuille's equation in the form,

$$\frac{dx}{dt} = \frac{R^2}{8\eta} \frac{dP}{dx} \quad (10)$$

and based on this equation states that it "follows that $R^2/8\eta$ is the conductivity ... of a specific pore". The original symbols are reported here: x is a linear coordinate, t is time, R is the pore radius, η is fluid viscosity, and P is pressure. Nutting also contributed to various experimental determinations of porosity and permeability during years when laboratory tests were not standardized. As far as the units of measures are concerned, he states (Nutting, 1927),

> "The rational definition of fluid permeability or conductivity (the difference is not clear, Editor's note) [of porous solids] is evidently the ratio of flow per unit area to the pressure gradient. In c.g.s. units, the specific conductivity is expressed in cc/second/cm^2 per dyne/cm^2/cm. This is the flow through a cube when the pressure difference between opposite faces is 1 dyne per cm^2."

This is the definition of hydraulic conductivity. In his paper, Nutting is still uncertain about the best method to measure permeability.

> "Probably the simplest of all methods of determining the fluid conductivity of porous solids is to measure the pressure drop along a cylindrical conductor along which fluid is moving at a given known rate. However, the experimental difficulties incident to determinations by this method are considerable and probably more accurate results can be obtained by the method above described."

The method embodies capillary rise, where permeability is calculated from a modified version of Eq. 10 as (original symbols),

$$-\frac{dh}{dt} = \frac{R^2 \sigma}{8\eta} \frac{\rho g}{l} (h - h_0) \tag{11}$$

where h is capillary rise, σ is effective pore space and l is length of the sand column. Actually, dh/dt is a velocity, and from the above "the conductivity of a porous solid for a fluid of viscosity η is $R^2\sigma/8\eta$, the factor ρg simply converts head into pressure units."

A few years later, in his important and much quoted paper of 1930, Nutting changed his mind and calculated the permeability by measuring the flow rate of a liquid through a cylindrical sample. He says,

> "The determination of permeability, ... is very simple in theory, but beset with minor difficulties. In principle, a plate of the rock to be tested is subjected to a known pressure difference and the rate of flow measured. From the dimensions of the test plate, the pressure gradient, viscosity, and rate of flow, the specific permeability is readily computed."

It is evident that here he refers to absolute permeability, as it is known today, but gives no formulas. He goes on to describe a standard apparatus used at the USGS to test permeability of oil sands: samples are 5 millimeters thick and about 0.5 inches in diameter. Then he specifies the units.

"The rational c.g.s. measure of permeability, namely, the flow in cubic centimeters per second through each square centimeter, of a fluid of unit viscosity under a pressure of 1 dyne per centimeter, is too small for convenience. A standard viscosity of 0.01 corresponding with water at 20 °C (68 °F) and a pressure gradient of 1 megadyne (1.02 kilograms weight) per centimeter, or 15.45 pounds per square inch per centimeter, is more convenient."

This is very close to the later definition of the darcy unit. Unfortunately, Nutting never cites the name or the experimental results published by Darcy.

In 1932 Tickell, Mechem and McCurdy of Stanford University proposed the following relationship to measure permeability of a cylindrical sample of rock: "Since there is viscous flow and constant viscosity Poiseuille's law holds and permeability is a constant proportionality factor in the equation (original symbols),

$$Q = K \frac{AP}{UL} \qquad (12)$$

where Q is discharge (cm^3/sec.), A is area of section (cm^2), L is length of section (cm), P is pressure applied (dynes/cm^2), and U is viscosity of fluids (poise). It would be convenient to adopt the term "perm" as the unit of permeability." This proposed unit definition (equivalent to 1 cm^2) did not have success in the scientific community of the time, and was never applied, but Eq. 12 is practically the modern form of "generalized" Darcy's law.

In 1933 Wyckoff, Botset, Muskat and Reed (all with Gulf Research & Development Corporation, Pittsburgh) published a paper entitled "The Measurement of the Permeability of Porous Media for Homogeneous Fluids" which in its very first page reports the modern definition of permeability and its physical assumptions:

"Permeability may be defined physically as the volume of a fluid of unit viscosity passing through, in unit time, a unit cross section of the porous medium under the influence of a unit pressure gradient, or, as the macroscopic velocity of a particle of a unit viscosity fluid at a point in the medium at which the pressure gradient is unity. Defined in this way, the permeability of a porous medium is independent of the absolute pressure or velocities within the flow system, or of the nature of the fluid, and is characteristic only of the structure of the medium. However, it has physical meaning only if the flow is of a viscous rather than turbulent character. [...] Analytically, the permeability, as well as the quantitative meaning of viscous flow, may be defined by means of Darcy's law, which may be stated as:

$$v_s = (k/\mu)(\partial p/\partial s) \qquad (13)$$

where v_s is the macroscopic velocity (fluid flux per unit area of porous medium) of the fluid, of viscosity μ, in the direction s at the point at which the pressure gradient is p/s and k is the permeability of the medium. If Eq. 1 is fulfilled with k having a constant value we shall consider the flow to be viscous. The empirical equivalent of Eq. 13, as applied to liquids, was established as long ago as 1856 by Darcy and has since been confirmed repeatedly by investigators here and abroad."

No theoretical justification was given for the above equation and no references made to earlier studies. Also, the equation was claimed not to be empirical in nature. However, Eq. 3 is the form of Darcy's law reported in the classic textbook of Muskat (1937), a sort of "Bible" for the many generations of Petroleum Engineers to come. Paramount in this study is the definition of the units in which permeability should be expressed. The dimension of permeability is area, but again Wyckoff *et al.* (1933) report that it "has been found convenient to adopt the following units: time: second; length: cm; volume: cc; pressure: atmos. (76.0 cm Hg); viscosity: centipoises. In view of Darcy's fundamental work in establishing the laws of flow for porous media it seemed appropriate to name the unit of permeability the darcy. We have then: k = 1 darcy ≈ 1(cc/sec.)/ cm^2/ (atmos./cm) for a fluid of unit viscosity". The darcy is still today the most used unit of permeability in petroleum engineering, and is equivalent to $0.987 \cdot 10^{-8}$ cm^2.

Apparently, before the rigorous experimental and theoretical formalization of Hubbert (1940, 1956, 1969), no researcher had clearly published the following experimental results that are obtained by varying one factor at a time in Eq. 12,

$$K \propto \rho, \quad K \propto \frac{1}{\mu}, \quad K \propto d^2 \tag{14}$$

where d is length, such as the mean grain diameter, which characterizes the size scale of the pore structure of the sand (Hubbert, 1956). Apparently Hubbert was the first to clearly express the above relationships. However, he strongly criticized Eq. 13, at least for its non-rigorous mathematical form, and thus influenced the acceptance of its general validity. In fact, Eq. 13 is expressed in terms of pressure gradient, so, strictly speaking, it is valid only for horizontal flow. Generalization of Darcy's law can be made only by considering the gradient of the potential function, defined as $\Phi = p + \rho gz$.

Conclusions

The permeability concept formulated by Darcy is known today as hydraulic conductivity, a flow parameter describing both the physical properties of the porous medium and the characteristics of the fluid. This paper presents some historical developments of permeability concepts. It reviews the earliest studies that contributed to the popularization of Darcy's Law which can still be found in many technical fields where the fluid is different than water. Particular in the petroleum industry, Darcy's Law was generalized for oil and gas flow, and the concept of multiphase flow was developed. This was accomplished by separating the influence of the rock from that of the liquid by reformulating the proportionality constant K. This paper tried to establish when and by whom the Darcy's Law was first generalized and made into a fundamental tool for applications in petroleum reservoir engineering. The origin of this generalization can be retraced in the work published by a number of American geoscientists at the end of the 1920s.

References

American Petroleum Institute (1935). *Standard procedure for determining permeability of porous media*, (ed. M. Muskat), Dallas - New York.

Amyx, J. W., Bass D. M. Jr., and Whiting, R.L. (1960). *Petroleum Reservoir Engineering*, McGraw-Hill, New York.

Barb, C. F. (1930). "Porosity-permeability relations in Appalachian oil sands." *Mineral Ind. Bull.*, 9:47-59.

Barb, C. F. and Branson, E. R. (1931). "Fluid flow through oil sands." *Intl. Petr. Tech.*, 8:325-334.

Botset, H.G. (1931). "Measurement of permeability of porous alundum discs for water and oil." *Rev. of Sci. Instruments*, 2: 84-95.

Brewster, M.A. (1925). "Discussion on some of the factors affecting well spacing." *Trans. AIME*, G-25:37-46.

Brown, G. O. (2002). "Henry Darcy and the making of a law." *Water Resources Research*, 38-7:1-12.

Darcy, H. (1856). "Détermination des lois d'écoulement de l'eau à travers le sable." Appendix, note D in *Les fontaines publiques de la ville de Dijon*, Victor Dalmont, Paris, 590-594.

Dupuit, A. J. (1863). *Etudes théoriques et pratiques sur le mouvement des eaux dans les canaux découverts et à travers les terrains perméables*, Dunod, Paris.

Fettke, C. R., and Copeland, W. A. (1931): "Permeability studies of Pennsylvania oil sands." *Trans. AIME*, 92: 329-339.

Fettke, C. R., and Mayne, R. D. (1930). "Permeability studies of Bradford sands." *Nat. Petr. News*, 22:61-62, 64-68, 107.

Forchheimer, D. H. (1898). "Grundwasserspiegel bei brunnenanlagen." *Zeit. Osterreichischeingenieur Architekten Ver.*, Wien, 50: 629-635, 645-648.

Freeze, R. A. (1994). "Henry Darcy and the fountains of Dijon." *Ground Water*, 32: 1, 23-30.

Hubbert M. K. (1940). "The theory of ground-water motion." *Jour. of Geology*, 48: 8, part 1.

Hubbert M. K. (1956). "Darcy's law and the field equations of the flow of underground fluids." *Trans. AIME*, 207:222-239.

Hubbert M. K. (1969). *The theory of ground-water motion and related paper*, Hafner, New York, 1969.

King, F. H. (1899). "Principles and conditions of the movement of ground water." *U.S. Geological Survey*, 19th Ann. Rept., Part 2, Denver, Colo., pp 59-294.

Meinzer, O. E. (1923). "Outline of ground water hydrology with definitions." *U.S. Geological Survey Water Supply Paper* 497, 71 pp.

Melcher, A. F (1920). "Determination of pore space of oil and gas sands." *Trans. AIME*, 65: (oil and gas), 469-497.

Melcher, A. F. (1925). "Apparatus for determining the absorption and permeability of oil and gas sands for certain liquids and gases under pressure.". *Amer. Ass. Petr. Geol. Bull.*, 9:442-450.

Mills, R. van A. (1921). "Relation of texture and bedding to movements of oil and water through sands." *Econ. Geol.*, 16:124-141.

Moore, T. V., Schilthuis, R. J., and Hurst, W. (1933). "Determination of permeability from field data." *API Prod. Bull.*, 211: 4-13.

Muskat, M. (1937). *Flow of homogeneous fluids through porous media*, McGraw-Hill, New York, 1937.

Muskat, M. and Botset, H. G. (1931). "Flow of gas through porous materials." *Physics*, 1:27-47.

Narasimhan, T. N. (1998). "Hydraulic characterization of aquifers, reservoir rocks, and soils: A history of ideas." *Water Res. Research*, 34:1, 33-46.

Nevin, C. M. (1932). "Permeability, its measurement and value." *Amer. Ass. Petr. Geol. Bull.*, 16:373-384.

Nutting, P. G. (1926). "Movements of fluids in porous solids." *Oil and Gas Jour.*, 26: 125.

Nutting, P. G. (1927). "The movements of fluids in porous solids." *Jour. Franklin Institute*, 313-324.

Nutting, P. G. (1929). "Some physical properties problems in oil recovery." *Oil and Gas Jour.*, 44: 160.

Nutting, P. G. (1930). "Physical analysis of oil sands." *Amer. Ass. Petr. Geol. Bull.*, 14:1337-1349.

Plummer, F. B., Woodward, J. S. (1936). "Experiments on flow of fluids through sands." *Trans. AIME*, 123:120-132.

Poiseuille, J. (1842). "Recherches expérimentales sur le mouvement des liquides dans les tubes de très petits diamètres." *Mém. savants étrangers, Comptes rendus de l'Académie des Sciences*, Vol. 9, pp. 433-534.

Ramazzini, B. (1692). *De fontium Mutinensium admiranda scaturigine, tractatus physico-hydrostaticus*, Modena.

Schriever, W. (1930). "Law of flow for the passage of a gas-free liquid through a spherical-grain sand." *Trans. AIME*, 86:329-336.

Slichter, S. C. (1899). "Theoretical investigations of the motion of ground water." *U.S. Geological Survey*, 19th Ann. Rept., Part 2, Denver, Colo., pp 295 ff.

Stearns, N. D. (1928). "Laboratory tests on physical properties of water bearing materials." *U.S. Geological Survey, Water Supply Paper* 596-F, 121-176.

Terzaghi, K. (1925). *Erdbaumechanic auf Bodenphysikalischer Grundlage*, Leipzig.

Thiem, G. (1906). *Hydrologische methoden*, Gebhardt, Leipzig.

Tickell, F. G. (1928). "Capillary phenomena as related to oil production." *Trans. AIME*, 82:343-361.

Tickell, F. G., Mechem, O. E. and McCurdy, R. C. (1933). "Some studies on the porosity and permeability of rocks." *Trans. AIME*, 103:250-260.

Vallisnieri, A. Sr, (1715). *Lezione accademica intorno l'origine delle fontane*, Venezia.

Wyckoff, R. D., Botset, H. G. and Muskat, M. (1932). "Flow of fluids through porous materials under the action of gravity." *Physics*, 3:90-113.

Wyckoff, R. D., Botset, H. G., Muskat, M. and Reed, D. W. (1933a). "The measurement of the permeability of porous media for homogeneous fluids." *Rev. of Sci. Instruments*, 4:394-405.

Wyckoff, R. D., Botset, H. G. and Muskat, M. (1933b). "Mechanics of porous flow applied to water-flooding problems." *Trans. AIME*, 103:219-249.

Wyckoff, R. D., Botset, H. G. and Muskat, M. (1934). "Measurement of permeability of porous media." *Amer. Ass. Petr. Geol. Bull.*,18:61-190.

Henry Bazin - Hydraulician

Willi H. Hager[1] and Corrado Gisonni[2]

Abstract

The recognition of Henry Bazin (1829-1917) as being one of the outstanding experimenters in hydraulics of the 19th century was founded by his 1865 report *Recherches Hydrauliques*. This paper presents the main findings of that historically important work including the fundamental results on uniform flow, open channel velocity distributions, gradually varied free surface profiles and waves. The latter are of particular interest because the solitary wave, previously observed by the Englishman Russell, was reexamined as well as wave breaking and surge formation due to a sudden blockage of flow. Other topics addressed are the story about the Darcy and Bazin uniform flow formula, and the reason why others that used the Bazin data sets produce formulas that are still in use. Engravings from the original 1865 report and photos from a recent visit to the canal site illustrate this note on a historically outstanding contribution to open channel hydraulics.

Introduction

Henry Bazin[§] (1829-1917) was probably the best and most famous open channel experimenter of the 19th century. Being introduced to the techniques of hydraulic observations by his mentor Henry Darcy (1803-1858), Bazin obtained a solid reputation for experimentation with his research on weir and jet flows. He set up an intense collaboration with the French mathematician Joseph Boussinesq (1842-1929), such that the two were at the forefront of hydraulics theory. Their results are still considered outstanding in the advance of hydraulic research. Whereas, Hager and Gisonni (2003) described Bazin's activities as a practicing civil engineer, this work highlights his research activities related to open channel flow as presented in Darcy and Bazin (1865). Additional background on Bazin is available in Bazin (1995) and Hager and Gisonni (2003).

[§] Contrary to most references, Bazin's birth registration, family vault and the Dijon memorial tablet, spell his first name as "Henry".

---

[1] Prof., VAW, ETH-Zentrum, CH-8092 Zurich, Switzerland
[2] Prof., Seconda Università degli Studi di Napoli, Via Roma 29, I-81031 Aversa (CE), Italy

Canal de Bourgogne Experiments

General. The Canal de Bourgogne is a prime inland navigation waterway that connects the North Sea with the Mediterranean. It has a total length of 242 km with a total of 189 sluices. Dijon, former capital of Burgundy, may be considered at its center of the canal, which was erected between 1826 and 1832. From 1840 to 1848, Henry Darcy was canal director while he served as Chief Engineer for the Côte d'Or Department. In 1854, Darcy asked to be relieved from normal duties because of poor health. He returned to Dijon and initiated a large series of experiments to investigate water flow in open channels. By that time, Henry Bazin had been transferred by the Corps to Dijon and posted to the canal service. After Darcy had lost two of his principal engineers, he started collaboration with Bazin, a young engineer with particular gifts in mathematics and accuracy. Although Bazin's sight was troubled, he would soon develop outstanding qualities in hydraulic experimentation.

Darcy passed away on January 3, 1858 while on a trip to Paris. Bazin became responsible for completing their common work, which took another seven years; partially for experimentation, but mainly for analysis and writing (Darcy and Bazin 1865). Figure 1 shows the title page of that work, (which the present authors were happy to find in a used bookstore in Dijon). The complete title is,

"Recherches Hydrauliques
Entreprises par M. H. Darcy, Inspecteur Général des Ponts et Chaussées
Continuées par M. H. Bazin, Ingénieur des Ponts et Chaussées
Première Partie
Recherches Expérimentales sur l'écoulement de l'eau Dans les Canaux Découverts
Deuxième Partie
Recherches Expérimentales sur la Propagation des Ondes"

This may be translated as "Hydraulic Research, Part 1: Experimental research on the water flow in open channels, Part 2: Experimental research on wave propagation." Noteworthy, an identical book was published as *Mémoires,* volume 19, authored by Bazin (1865) alone. The Darcy and Bazin report also contains a review on the 'paper', submitted by Dupin, Poncelet, Combes, Clapeyron and Morin, the latter chairing that committee during the meetings on July 27 and August 3, 1863 (Clapeyron 1862a). Both Morin and Poncelet were outstanding hydraulic engineers and the committee did an excellent job. Their report covers 24 pages and states that Bazin had attempted to explain Darcy's findings relative to pipe flows, where the effect of wall roughness was found to be significant. Bazin, when taking over works from his predecessor in 1856, had to complete a large project, and the reviewers stated that the merits for the work should be directed toward Bazin. Bazin (1863) published a pre-report once the main data analysis was completed, which was also translated in Italian by Pareto (1864). Bazin was awarded the *Prix Dalmont* from the *Académie des Sciences,* Paris, which carried with it 3,000 French francs, at that time a considerable sum of money (Prix Dalmont, 1868).

Report Structure. The book is separated in four chapters; Experiments on uniform flow, Experiments on the velocity distribution, Experiments on gradually varied flow, and Wave movement in open channels. A total of 652 pages are accompanied with 33

MÉMOIRES

PRÉSENTÉS PAR DIVERS SAVANTS

À L'ACADÉMIE DES SCIENCES

DE L'INSTITUT IMPÉRIAL DE FRANCE

ET IMPRIMÉS PAR SON ORDRE

SCIENCES MATHEMATIQUES ET PHYSIQUES

TOME DIX-NEUVIÈME

PARIS

IMPRIMERIE IMPÉRIALE

M DCCC LXV

Figure 1. Title page of 1865 report (Darcy and Bazin 1865)

plates. Uniform flow experiments were conducted between 1855 and 1862 on an artificial channel along *bief* 57, (reach 57) of the *Canal de Bourgogne,* just south of Dijon. That reach has a length of 596.5 m. The channel was parallel to the Canal over the first 450 m and then took a left to divert waters into the Ouche River (Figure 2). With a width of 2 m and a depth of 0.95 m, it was designed such that the wall roughness and channel slope could be varied to the conditions of experimental interest. Water was supplied from the canal and gates at a special intake basin controlled by the discharge.

Darcy's interest was not only uniform flow, but also velocity distributions in free surface flows, given that these were not systematically analyzed until 1850. Darcy realized that the Pitot tube was better suited for exact experiments than propeller

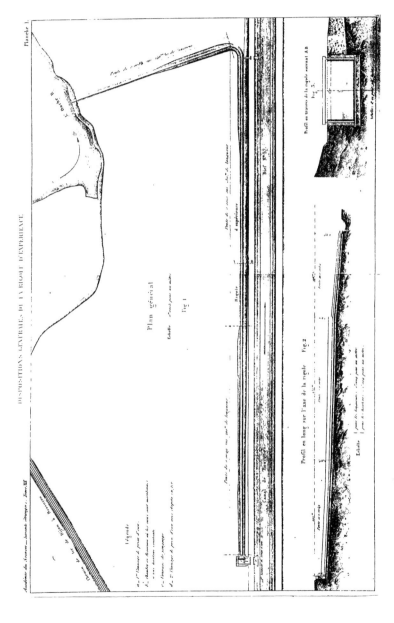

Figure 2. Channel plan, longitudinal and transverse sections [Plate I, Part 1]

meters, as proposed by Woltman and later improved by André Baumgarten (1808-1859), one of his assistants prior to Bazin's arrival. Bazin tested the improved Pitot-Darcy tubes by, (1) comparison with surface floats, (2) towing in still water, and (3) discharge measurement by integration over a section. Procedures (1) and (3) resulted in deviations of less than 1% from observations, whereas (2) had 3.4% deviation mainly due to a suspension problem. The average of the other two procedures gave a coefficient 1.000 for the Pitot-Darcy tube, thus excellent agreement with a basic hydraulic principle. Free surface elevations were measured with point gages, and stilling wells were used for static water levels (Figure 3). The following sections highlight the main findings.

Uniform Flow. Consider an open channel whose slope I, hydraulic radius R, and wall roughness are constant. For a given discharge Q, conditions may be establish where the flow depth along the channel reach remains constant, velocity distribution from one position to the next do not vary. Thus, equilibrium occurs between the tractive and resisting forces. The constant flow depth will correspond to the uniform flow depth. Such idealized flow was successfully considered since the 18th century. Antoine de Chézy (1718-1798) proposed a proportionality for the average cross-sectional velocity U, which was extended to the non-homogeneous equation,

$$bU^2 = RI \qquad (1)$$

where b is a dimensional "roughness" coefficient [s^2/m]. Gaspard-Marie Riche de Prony (1755-1839) proposed expanding (1) with the addition of a linear velocity term aU on the left side. The effect of that term was known to be small, except for so-called laminar flows investigated by Poiseuille and Darcy in the 1840s and 1850s. Darcy's interest was thus directed to that anomaly. Why would certain flows follow a linear velocity law and others a quadratic law? Where was the transition between these two basic types of flows? What were the parameters that describe fluid flows in general? Finally, what were the differences between pressurized and free surface flows?

Another point of interest was the effect of wall roughness. Up to Darcy's pipe flow experiments that effect was not systematically considered. Darcy (1857) found that wall friction, or the nature of wall surface, had a significant effect on the pipe discharge and velocity distributions. Thus, the *Canal de Bourgogne* research project was directed to shed light on that question in open channel flow. In 1855, Baumgarten had conducted preliminary experiments on the *Roquafaveur* aqueduct close to Marseille and verified Darcy's presumption of a significant wall effect. Experiments conducted by Bazin in 1856 on the *Canal de Bourgogne* site involved four wall types: pure cement, brick and stones, wooden planks and gravels with diameters of 0.01 to 0.02 m, and 0.03 to 0.04 m (Figure 4). The ratio RI/U^2 as given by (1) varied slightly with increasing discharges and a given wall roughness, but changed significantly for different wall roughness. Therefore, the wall roughness was shown to have a definite effect on fluid flow, a major conclusion of the 1865 report (Figure 5).

Bazin, in 1858 and 1859, investigated the effect of bottom slope on the velocity by using wooden planks arranged in the channel. Slopes of 0.15%, 0.59% and 0.886% were tested for discharges between 0.10 and 1.24 m^3/s. It was found that the slope effect is perfectly accounted for in (1). Today, one would state that the quadratic friction law applies for fully turbulent, rough flows. Next, the effect of cross section

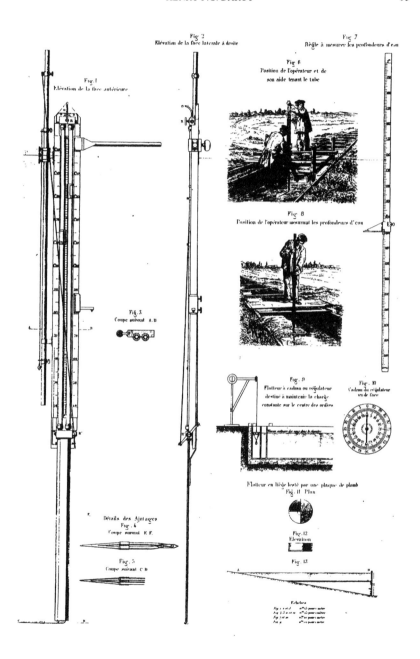

Figure 3. Instrumentation used in *Canal de Bourgogne* experiments [Plate IV, Part 1]

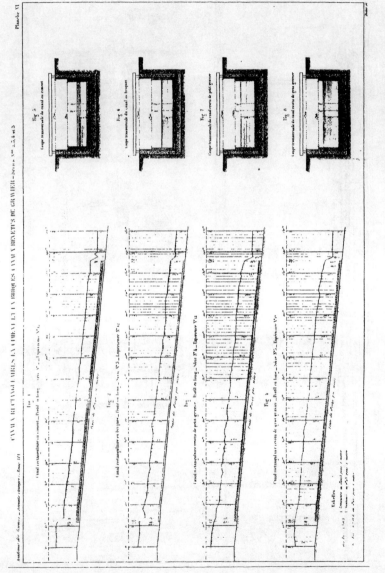

Figure 4. Typical 'uniform' flow (left) and types of wall roughness investigated (right) [Plate VI, Part 1]

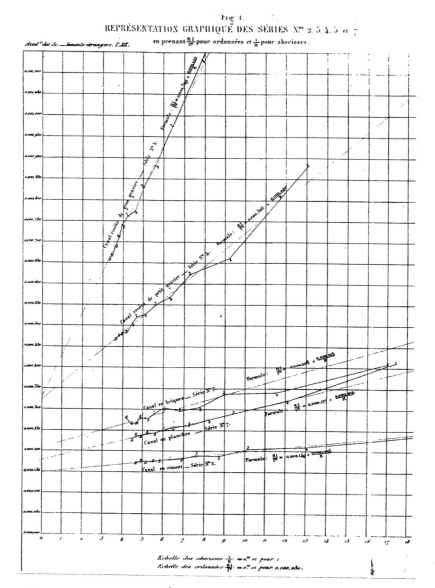

Figure 5. Plot $1/U$ versus RI/U^2 for Series 2, 3, 4, 5 and 7 [Plate VIII, left, Part 1]

shape was investigated by examining rectangular, trapezoidal, triangular and semi-circular channels. It was concluded that the shape effect was also perfectly accounted for in (1), provided the constant b incorporated the hydraulic radius R. Finally, the results were plotted as Darcy did for his pipe flow observations following,

$$RI/U^2 = \alpha + \beta/R = \alpha(1 + \beta_1/R) \qquad (2)$$

where the coefficients α and β or β_1 depend exclusively on the wall roughness pattern. Bazin proposed four roughness categories,

1. very smooth wall $\alpha = 0.00015$ $\beta_1 = 0.03$
2. smooth wall $\alpha = 0.00019$ $\beta_1 = 0.07$
3. not very smooth wall $\alpha = 0.00024$ $\beta_1 = 0.25$
4. earthen wall $\alpha = 0.00028$ $\beta_1 = 1.25$

The four categories were described with typical examples from practice, but not quantified as is now done with the wall roughness. (That issue was only resolved 60 years later by the experiments conducted at Göttingen, Germany.) Bazin was certainly not the most experienced engineering researcher in 1865. Thus, he tested the simplest approach by following his master's equation and verified basic agreement between free surface and pressurized flows. The introduction of categories of wall roughness had been proposed by Darcy earlier and was generally adopted. Only a rough picture of reality was obtained by the four categories since there existed transition cases, and the result would be significantly different if one or the next category was used.

A significant observation applies to the question of dimensional homogeneity. Dimensional analysis of the left hand side of (2) shows that α has dimensions of second squared per meter. By including the gravitational acceleration g as gRI/U^2, α becomes dimensionless. This detail was certainly known by Darcy and Bazin, but not considered important. Remember dimensional theory did not yet exist and the inclusion of g would require an additional computational step. More importantly, β_1 has the dimensions of meters. By inspecting the numbers quoted, it may be realized that β_1 indeed increases as the roughness (length) increases.

The disadvantage of (2) is the need of two coefficients α and β. Others heavily criticized that with the proposal of including only one coefficient, such as $1/n$ in the Gauckler-Manning-Strickler equation (Hager 1994). To end this story, Bazin (1871) was not ready to follow that argument, and replace his and Darcy's formula with another until he would replace (2) by the simpler expression (Bazin 1897)

$$(RI)^{1/2}/U = (1 + \gamma/R^{1/2})/87 \qquad (3)$$

with γ as a unique roughness coefficient. It thus took Bazin more than 30 years until he accepted a single coefficient. For him it was certainly an issue of faithfulness towards his mentor. Darcy had proposed a different formulation, and as his protégé, Bazin wished to respect it.

Equation (2) was also tested using another series of experiments conducted at the *Grosbois* site (Hager and Gisonni 2003) with a trapezoidal channel of slope of 1:10 (horizontal to vertical) and an almost smooth boundary. Average velocities ranged from 2.75 to 6.43 m/s, with maximum surface velocities up to 10 m/s. These flows were certainly aerated, such that observations were complicated and the accuracy reduced as compared with the Dijon experiments. The agreement of the two series was reported as poor, without further conclusions. Bazin also compared his observations with others conducted mainly in the 1850s in France, by accounting for the wall roughness, and found satisfactory agreement.

Velocity characteristics. Once (2) had been established, Bazin continued with the characteristics of the velocity distribution in open channel flows. Of practical significance was the ratio between the average velocity U and the maximum velocity V, the latter having often been observed with floats. Bazin first stated that floats normally do not have a velocity that corresponds to the maximum value because the velocity varies in the vertical direction, from a surface value to a maximum that is located somehow lower, and then to a much smaller bottom value. The topic was tackled with a special paper by Bazin (1884). Darcy and Bazin, as well as Boussinesq (1877), adopted a *finite* wall velocity because the concept of wall turbulence was not yet advanced, and the instrumentation was unable to measure the velocity distribution in the boundary layer.

Bazin noted that the ratio of velocities V/U was not constant but varied with the parameter (RI/U^2), as used in (2). By plotting the results of observations with these parameters, he realized that simply a factor $K = 14$ may be introduced and that the maximum velocity V is then given by

$$V/U = 1 + K(RI/U^2)^{1/2} \qquad (4)$$

Bazin presented contour plots for open channel velocity distribution, certainly the first available and the basis for future studies in internal hydraulics. These plots clearly demonstrate that the velocity distribution is more uniform for flows in channels having smooth boundaries than with large wall roughness, in agreement with (4). Figure 6 shows typical results for velocity contours observed in various channel geometries having wooden walls.

Bazin also investigated the effect of air resistance on open channel flow. In 1858 and 1859, he designed two rectangular ducts of widths 0.80 m and 0.48 m with heights of 0.50 m and 0.30 m, respectively. In a first series, flow was measured with the duct enclosed, but the water not filling the section. Then the cover was removed and flow measured in the open channel arrangement. The difference was so small that no specific effect could be attributed to ducted flow as compared with the corresponding channel flow. Accordingly, Bazin's velocity formula may also be applied for sewers, that is, channels that are covered with a limited access of air. Bazin also investigated pressurized flow and compared it with corresponding free surface flows, for which a definite difference was observed. For pressurized flow, the boundary layer produces a zero velocity along the entire pipe perimeter, whereas the free surface velocity is close to the average velocity for free surface flow. Bazin noted that the velocity distribution of free surface flows is much more complex than of pressurized flow. For pressurized circular pipe flow, there is a full symmetry and the velocity distribution may be expressed only as a function of radius. In contrast, for partially filled pipe flow, such symmetry is lost and the distribution is a function of radius *and* angle from the center (Figure 7).

Bazin even analyzed the vertical velocity distribution $v(h)$ in rectangular channels and developed a simple formula. With H as flow depth, v as variable velocity and h as vertical coordinate, he found,

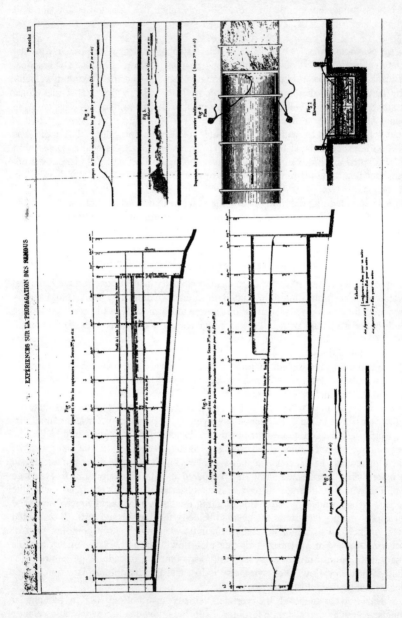

Figure 6. Velocity contours for channels with wooden walls (Series 67 to 70) [Plate XXI, 1]

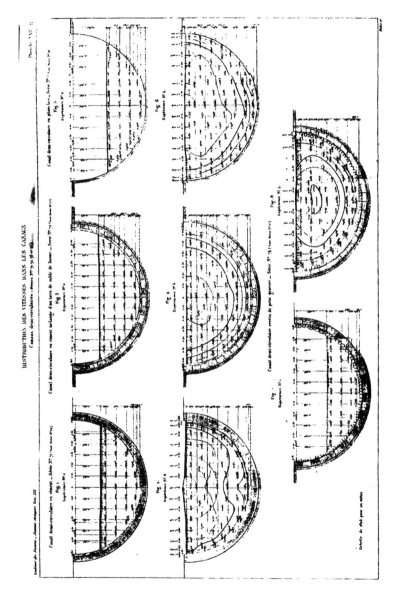

Figure 7. Velocity distribution in semi-circular open channel, for various boundary roughness [Plate XXII, Part 1]

$$V - v = K(RI)^{1/2}(h/H)^2 \tag{5}$$

where V is the surface velocity and K a variable involving only wall roughness (Figure 8). These observations were made with only small flow depths between 0.084 and 0.38 m, such that (5) was rather sensitive to small inaccuracies. For $h = H$, the bottom velocity is given as $v_b = V - K(RI)^{1/2}$, while at the surface $v_s = V$. Equation (5) was the basis of theoretical research later conducted by Joseph Boussinesq (1842-1929). That formulation, extended with the gravitational acceleration g, was then used in the 20th century for boundary layer analysis. The final forms as presented by Ludwig Prandtl (1875-1953) and Theodor von Karman (1881-1963) were thus founded on the concept of Darcy and Bazin, thereby accounting for hydraulic similitude involving the Reynolds number.

Gradually varied flow. The third section of Bazin's work addressed steady, non-uniform flow, in which the depth may vary from section to section. These flows were extensively discussed by Joseph Bélanger (1790-1874) and Jean-Victor Poncelet (1788-1867), the latter being a member of the review committee. Bazin was able to demonstrate that the so-called gradually varied flow equation may be applied by inserting an appropriate friction slope. These flows present a particularity for transitions from supercritical to subcritical regimes, i.e. the hydraulic jump. Open channel flows are currently described by the ratio of the velocity to the wave celerity, given by Froude number. However, both Adhémar Barré de Saint-Venant (1797-1886) and Joseph Boussinesq (1842-1929) realized earlier the significance of the parameter $U/(gh)^{1/2}$. Bazin conducted several experiments involving hydraulic jumps, but only with Froude numbers between 1 and 2. Thus, only undular jumps occurred. These wavy transition phenomena dissipate much less energy than direct jumps characterized by air entrainment, large turbulence production and significant energy dissipation (Figure 9). In contrast to the momentum approach of Bélanger, Bazin's computations were incorrect because he assumed no energy loss across the jump. At the end of that section, the review committee stated,

"By following the research initiated by Darcy, and by completing it with a scientific and lucid discussion, Mr. Bazin has not only made a work of great value for the engineering profession motivated by the souvenir to his former head, he (Bazin) also acknowledged the support of Darcy in terms of conception and general organization of the work, but his (Bazin's) personal contribution is significant and one cannot appreciate enough the honor towards the *Académie des Sciences* and his corps (*Ponts et Chaussées*). As a consequence, Your commissioners propose to the academy of sciences to accept the memoir of Mr. Bazin, and ask to publish it in the *Recueil des savants étrangers*, in addition to this report to the minister of agriculture, commerce and public works. The conclusions of that report are accepted."

Wave experiments. The Englishman John Scott Russell (1808-1882) published two reports on waves (Russell 1837 and 1845) in which he described a new type of wave that was referred to as the solitary wave. Instead of a usual sinusoidal wave with peaks and troughs, a solitary wave has a single peak extending to the still water elevation. They may be produced with a piston at the end of a channel that suitably displaces the

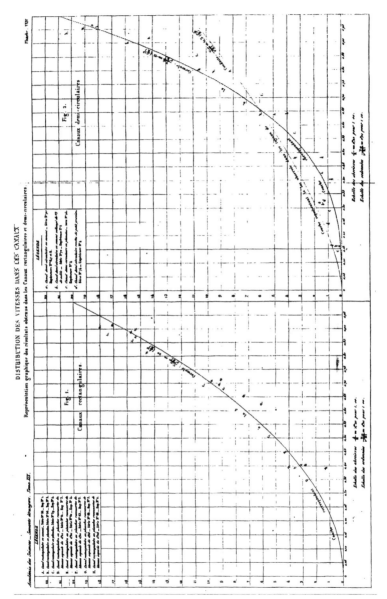

Figure 8. Velocity distribution in a semi-circular channel, observations compared with (5) [Plate XXIV, Part 1]

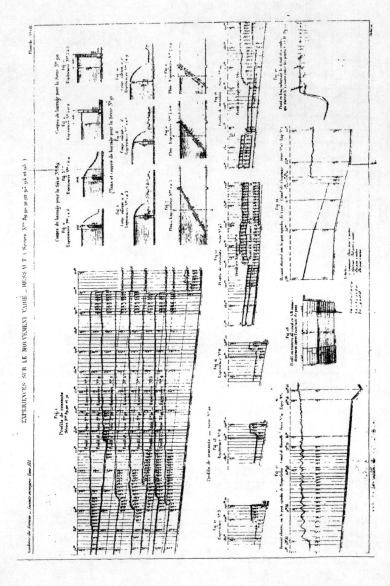

Figure 9. Weak hydraulic jumps with backwater curves up- and downstream, and details of weir overflows [Plate XXVIII, Part 1]

water body. The question whether a negative solitary wave may be produced by moving the piston back from the water body was also tested, with the result of a negative wave train. Accordingly, only solitary waves with a positive wave crest are physically amenable. Russell determined the propagation velocity, c of a solitary wave experimentally as,

$$c = [g(H + h)]^{1/2} \qquad (6)$$

where H is the still water depth and h the height of the wave crest. Bazin (1865) stated that Russell's experiments were conducted in a small channel, and that his own observations on the larger channel fully confirm (6). Bazin extended his research to the case where a solitary wave was produced in a channel flow of velocity U, and found that the propagation velocity simply is

$$c = U \pm [g(H + h)]^{1/2} \qquad (7)$$

Accordingly, a moving observer of relative velocity $c - U$ would see wave propagation to either side of the disturbance, in analogy to an elementary wave that propagates according to (6) as $\pm(gH)^{1/2}$. Bazin stated that results in a channel with a finite velocity U were less accurate than those in still water, simply because of the disturbances of the flowing water.

Bazin then investigated three cases of practical interest, using a slightly modified installation also along *Canal de Bourgogne* (Figure 10):

1. Water is added to a channel containing still water, how does that wave propagate? Bazin realized the significant effect of relative wave height. For small waves, a continuous wave train may establish, for large relative wave height, the first wave crests break. These conditions may be realized in a sloping channel in which the water is initially at rest, as seen in the center plot of Figure 11. A criterion for wave breaking was also given that can actually be considered an upper limit. Also, the initial still water depth undergoes variations and the wave celerity was observed to be $c = (gH)^{1/2} + (3/5)U$, where U is the velocity given by the unit discharge added, divided by H.
2. In a channel with a constant discharge, the flow is suddenly stopped at the downstream end by positioning a gate. What is the effect of that abrupt change? Such experiments were previously conducted by Giorgio Bidone (1781-1839) and Darcy. Because the water flow in the channel does not abruptly stop, there is an increase of flow depth close to the gate resulting in a surge wave moving against the flow (Figure 12). In that case, the wave celerity may be predicted with (7) using the negative sign, or with $c = (gH)^{1/2} - (2/5)U$. Wave breaking occurred, provided the channel velocity U is larger than $(1/2)(gH)^{1/2}$.
3. If water is discharged from a side channel for the same initial conditions as in 2, then waves are set up that can be compared with the action of the sea on maritime rivers. These observations were initiated by Darcy in 1857 and completed in 1859 by Bazin. However, the flows were so complicated that a detailed description was beyond the scope of the paper

The wave experiments were made over a length of 432 m with two gates set at the ends to control the boundary conditions. Waves were observed with 13 gages spaced at 20 m. A reach for wave development was added upstream of the first wave

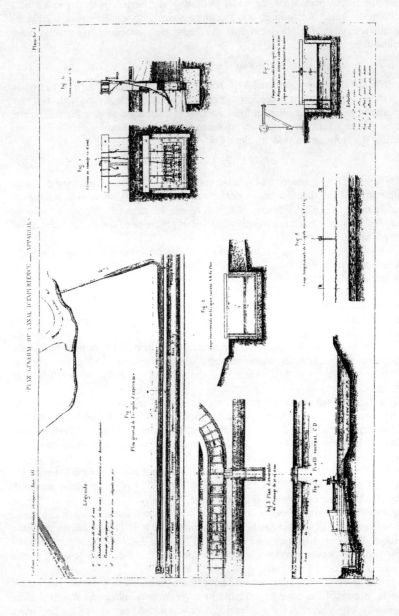

Figure 10. Arrangement for wave observations, including gates (right) for abrupt discharge variation, and wave gages [Plate I, Part 2]

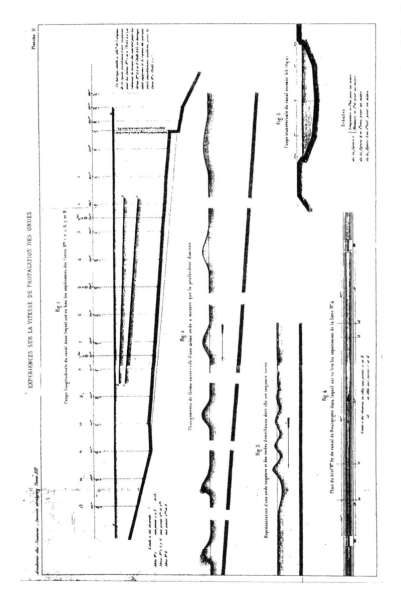

Figure 11. Test arrangements for wave propagation tests, with positive (center) and negative (below) waves [Plate II, Part 2]

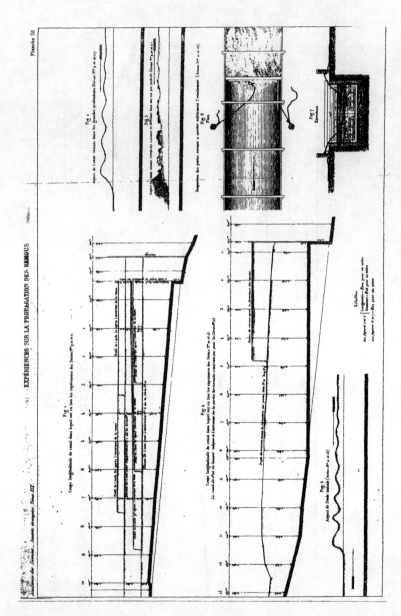

Figure 12. Observations of surge against flow direction, either continuous or breaking [Plate III, Part 2]

gage. All gages were originally positioned to the still water level and readings were made by individual assistants as the wave passed. In total, 11 experiments for case 1 were conducted, including one with an adversely sloping bed in the direction of wave propagation. That experiment may be compared with wave run-up at a shore where an originally smooth solitary wave may become so steep that wave breaking occurs (Figures 11 and 12). It was also observed that the celerity of a wave, having once broken, is greatly reduced compared with a usual wave, because of energy dissipation.

Negative waves were stated to be much more difficult to observe because of a gradual variation of the free surface profile. Fifteen experiments in total were conducted. Negative waves always induced a wave train, and the wave height was dramatically reduced over a relatively small reach. Additional observations related to the reflection times of waves, and up to seven cycles back and forth were recorded. These times remained practically constant, because of the extremely small energy dissipation involved. The entire wave report with more than 150 pages and 5 plates covers approximately a fourth of the 1865 paper. A review report is also available as the first part of the 1865 report by Clapeyron (1862b).

Shortly after Bazin's completion of works on water waves, two important additions were published. Saint-Venant (1871) presented the one-dimensional equations of unsteady open channel flows, based on the assumption of hydrostatic pressure and uniform velocity distributions. Boussinesq (1871) extended that approach by accounting for non-hydrostatic pressure distribution, and thus provided the first theoretical account on solitary waves. A thorough report was published in *Mémoires*, which marked an important addition to hydraulics in the 19th century (Boussinesq, 1877). Later, Bazin and Boussinesq would strongly collaborate on weir flow. Accordingly, the best experimenter and hydrodynamicist of France had found a suitable collaboration that would have a strong impact on the 20th century hydraulics. Wave hydraulics has not been systematically reconsidered until the 1930s, given the complexity of the phenomena and the difficulties with instrumentation. That also points to the excellence of Bazin's research and the importance of his results.

Impact of Bazin's Research

In the Introduction to *Recherches Hydrauliques*, Bazin (1865) stated that it was a great honor to finish a work that was initiated by his former head Henry Darcy, and that he intended to do the best for the reputation of the two. Bazin certainly had a reputable name after the work was published. However, it can be questioned, if Darcy would really have dreamt of such a large and outstanding work. Of course, Darcy had published two large research projects, *Les fontaines publiques* (Darcy 1856) and *Recherches expérimentales relatives au mouvement de l'eau dans les tuyaux* (Darcy 1857). The first work was directed to practitioners and addressed details for water supply engineering, including topographical, hydrological and material questions, as well as how fountains may be installed and how pipes can be fabricated. The particularity of that work is a small Appendix in which the basic groundwater law was experimentally determined. That addition pretends that *Les fontaines publiques* is a research work, but it was definitely not. In contrast, the pipe flow report was definitely a research work, which partially spoke to Darcy's interest in these flows. As mentioned in *Les fontaines publiques*, there were always design problems for water mains because discharges were poorly predicted. Once at Paris, Darcy had the staff, the installation and governmental sponsoring for that research. He carried out an impressive project. Nothing comparable had been presented before and his data was employed for decades. As any hydraulic engineer knows, pipe flows are simpler to measure than free surface flows. Bazin thus presented an even larger work with a staff and financial support that seems to have been similar to Darcy's pipe research at Paris.

There is one point to discuss. Why would Bazin adopt his master's flow formula? There have been dozens of proposals for uniform open channel velocity, and Bazin could easily present a different equation. As mentioned, Bazin considered his work together *with* Darcy, and he felt to do right towards his former colleague to adopt, or at least to check, the Darcy pipe flow formula for open channel conditions. It fitted well, as previously discussed, such that there was no reason to look for something different. However, Bazin added considerably to Darcy's original project with his sections on velocity distribution (similar to that made by Darcy in the pipe flow report, but to a much smaller extent), on backwater curves that had not received systematic attention up to then, and his wave experiments. Bazin's data sets are still in use, they were the basis of many research projects in the early 20th century, such as of Heinrich Blasius (1883-1970), Johann Nikuradse (1894-1979) and the American Ralph Powell (1889-1975). As described by Dooge (1987) and Hager (1994), there was much confusion in open channel hydraulics because of the mass of data, which was often poor quality. Bazin's data sets were considered extremely reliable. Thus he not only completed an original research project relating to velocity distribution and wave hydraulics, but also added significantly to hydraulic modeling. His channel arrangements, his instrumentation and his computational approaches were an example for later research. Thus, it was this threefold impact methodology, analysis and originality that makes up the excellence of Bazin's 1865 research.

Apart from the 1863 paper and his 1865 report, Bazin did not publish any engineering paper on that research until 1871, being urged by the widespread use of his data sets to state again his opinion. Starting in 1867, various individuals had successfully used the Bazin data to present even simpler formulas for uniform channel flow. Mention might be made of Lévy (1867), and Gauckler (1867, 1868), which culminated in the Gauckler-Manning-Strickler GMS-formula that is currently still in extensive use (Hager 2001). Also of note were Hagen (1868, 1869) known for the laminar pipe flow experiments and the Poiseuille-Hagen formula. Kutter (1868) and Ganguillet and Kutter (1869) proposed a complex formula that covered all the data sets of Humphreys and Abbot and Darcy and Bazin. Likewise, Grashof (1869) known for the Grashof number in hydraulics, Regis (1869) with a graphical solution for various formula, and Dal Bosco (1871) known for the first translation of the Ganguillet-Kutter paper followed. Bazin (1871) thus reviewed various of these proposals by subdividing into power formulas those of de Saint-Venant (1851), Gauckler (1868), Hagen (1868) and Bornemann (1869), into binomic formulas those of Prony, Eytelwein and Darcy and Bazin (1865), and even more complicated approaches those of Humphreys and Abbot (1861) and Ganguillet and Kutter (1869). It was stated that the data of Humphreys and Abbot were in disagreement with all European data sets, certainly because of different methods of observation. Accordingly, also the Ganguillet and Kutter approach was in error, because one of its important foundations was the

Figure 13. *Canal de Bourgogne* at *bief* 57 in 2001

American data. Bazin further criticized the Gauckler and the de Saint-Venant approaches, the first because of an artificial subdivision into two formulas without physical reason, the second because of a relatively poor data basis. Bazin, at that time and in the journal of his Corps, was fully convinced of his research analysis, regardless of the obvious disadvantages of (2) with two parameters describing the same effect.

Bazin's Remnants

The present authors visited Dijon twice, once in 2001 and then in 2002, the latter trip devoted to topics other than the 1865 research. A report on the 2001 trip is available without much note on the *Canal de Bourgogne* experiments (Hager and Gisonni 2002). In contrast to the conclusions of another paper, evidence of Darcy and Bazin's presence can still be seen at Dijon, their works being excellently documented by structures that have persisted more than 150 years, including a particularity beautiful *Place Darcy*. The documentation of the 1865 report is maybe the most difficult, because the channel was filled and has disappeared. The second channel used in the 1880s and 1890s for weir research may be discovered under a huge plantation of bush. *Canal de Bourgogne* including *bief 57* and *l'Ouche* are still there, currently less polluted than 150 years ago but nothing exciting for tourists (Figure 13). Other written remnants can be found at the municipal library of Dijon, which is also the seat of Dijon Academy, of which Bazin was a member since 1867, and where his papers are kept.

There was a particular point that brought us to Dijon, instrumentation. We explored whether the Pitot-Darcy tube was still to find, whether some gages could be observed or if maybe other material used during this experimental campaign was still there. So far, we have been unable to locate any instrumentation. Only the small

Figure 14. Canal house, *bief* 57 in 2001

houses close to the *Canal* are in their original condition. They are used for lock operation and are rented by families during summer holidays. Figure 14 shows the *bief 57* house. The best account of Bazin research remains *Recherches hydrauliques*.

Conclusions

One of the outstanding hydraulic works of the 19th century was conceived, executed and written by Henry Bazin, former assistant and then colleague of his predecessor Henry Darcy. The main findings of that work were highlighted in their paper, together with a description of the environment in which that research was conducted. The 1865 report motivated a number of hydraulicians to analyze the data, resulting in equations that are still in use. Bazin's close relation and bond to Darcy is why Bazin did not attempt a simplification of his uniform flow formula until 1897. Bazin's research provided a significant advance in hydraulics. It is less the formula, but much more the experimental procedures and results by which his name has survived in hydraulics.

REFERENCES

Bazin, G. (1995). *Henry Emile Bazin*. Private printing, Annemasse.

Bazin, H. (1863). "Expériences sur les lois de l'écoulement de l'eau dans les canaux". *Le Technologiste,* 24:91-93; 25:93-98; 25:157-159; 25:210-211; 25:322-325; 25:385-388.

Bazin, H. (1865). "Recherches hydrauliques". *Mémoires* présentés par divers savants à l'Académie des Sciences de l'Institut Impérial de France 19:1-652.

Bazin, H. (1871). "Etude comparative des formules nouvellement proposées pour calculer le débit des canaux découverts". *Annales des Ponts et Chaussées* 41(1):9-43.

Bazin, H. (1884). "Notice sur l'emploi des doubles flotteurs pour la mesure des vitesses dans les grands cours d'eau". *Annales des Ponts et Chaussées* 54(1):554-591.

Bazin, H. (1897). "Etude d'une nouvelle formule pour calculer le débit des canaux découverts". *Annales des Ponts et Chaussées* 67(4):20-70.

Bornemann, K.R. (1869). "Die Gauckler'sche Theorie der Bewegung des Wassers in Flüssen und Canälen". *Der Civilingenieur* 15:14-52.

Boussinesq, J. (1871). "Théorie de l'intumescence liquide appelée onde solitaire ou de translation, se propageant dans un canal rectangulaire". *Comptes Rendues de l'Académie des Sciences,* Paris 72:755-759.

Boussinesq, J. V. (1877). "Essais sur les eaux courantes". *Mémoires* présentés par divers savants à l'Académie des Sciences de l'Institut de France 23:1-680; 24:1-60.

Clapeyron, E. (1862a). "Expériences sur les lois de l'écoulement de l'eau dans les canaux découverts". *Comptes Rendus de l'Académie des Sciences,* Paris 55:274-277; 57:192-205; 57:255-264.

Clapeyron, E. (1862b). "Expériences sur les ondes et la propagation des remous". *Comptes Rendus de l'Académie des Sciences,* Paris 55:353-357; 57:302-312.

Dal Bosco, B. (1871). "Cenni generali storici sulle nuove formole sperimentali del moto dell'acqua entro canali ed alvei sistemati di fiumi e sulle memorie degli Ingg. E. Ganguillet e W.R. Kutter intorno a questo argomento". *Giornale dell'Ingegnere Architetto* 19:441-452.

Darcy, H. (1856). *Les fontaines publiques de la ville de Dijon*. Dalmont, Paris.

Darcy, H. (1857). *Recherches expérimentales relatives au mouvement de l'eau dans les tuyaux*. Mallet-Bachelier, Paris.

Darcy, H. and Bazin, H. (1865). *Recherches hydrauliques*. Imprimerie Impériale, Paris.

Dooge, J.C.I. (1987). "Historical development of concepts in open channel flow". *Hydraulics and hydraulic research*, 205-230, G. Garbrecht, ed. Balkema, Rotterdam.

Ganguillet, E. and Kutter, W.R. (1869). "Versuch zur Aufstellung einer neuen allgemeinen Formel für die gleichförmige Bewegung des Wassers in Canälen und Flüssen, gestützt auf die Resultate der in Frankreich vorgenommenen umfangreichen und sorgfältigen Untersuchungen und der in Nordamerika ausgeführten grossartigen Strommessungen". *Zeitschrift des Österreichischen Ingenieur- und Architekten-Vereins* 21:6-25; 21:46-59.

Gauckler, P. (1867). "Etudes théoriques et pratiques sur l'écoulement et le mouvement des eaux". *Comptes Rendus de l'Académie des Sciences*, Paris 64:818-822.

Gauckler, P. (1868). "Du mouvement de l'eau dans les conduites". *Annales des Ponts et Chaussées* 38(1):229-281.

Grashof, F. (1869). "Humphreys' and Abbot's Theorie der Bewegung des Wassers in Flüssen und Canälen". *Zeitschrift des Vereines deutscher Ingenieure* 13(5):289-300; 13(6):353-364; 13(8):482-489.

Hagen, G. (1868). "Über die Bewegung des Wassers in Strömen". *Mathematische Abhandlungen der Königlichen Akademie der Wissenschaften* Berlin 24(2):1-29.

Hagen, G. (1869). "Über die Bewegung des Wassers in cylindrischen, nahe horizontalen Leitungen". *Mathematische Abhandlungen der Königlichen Akademie der Wissenschaften* Berlin 26:1-29.

Hager, W.H. (1994). "Die historische Entwicklung der Fliessformel". *Schweizer Ingenieur und Architekt* 112(9):123-133.

Hager, W.H. (2001). "Gauckler and the GMS formula". *Journal of Hydraulic Engineering* 127(8):635-638.

Hager, W.H. and Gisonni, C. (2002). "Finding Darcy at Dijon". *Journal of Hydraulic Engineering* 128(5):454-459.

Hager, W.H. and Gisonni, C. (2003). "Henry Bazin – Civil engineer". *Journal of Hydraulic Engineering* 129(3):171-175.

Humphreys, A.A. and Abbot, H.L. (1861). *Report upon the physics and hydraulics of the Mississippi River*. Government Printing Office, Washington.

Kutter, W.R. (1868). "Die neue amerikanische Theorie der Bewegung des Wassers in Flüssen und Canälen von Humphreys und Abbot und die bisherigen europäischen Formeln". *Cultur-Ingenieur* 1:293-310.

Lévy, M. (1867). "Théorie d'un courant liquide à filets rectilignes et parallèles de forme transversale quelconque. Application aux tuyaux de conduite". *Annales des Ponts et Chaussées* 37(1):237-319.

Pareto, R. (1864). "Movimento dell'acqua nei canali scoperti". *Giornale dell' Ingegnere-Architetto* 12:155-161; 12:496-502.

"Prix Dalmont". (1868). *Comptes Rendus de l'Académie des Sciences,* Paris 66:935-937.

Regis, D. (1869). "Tavole grafiche a due argomenti fatte sulle formole di Darcy e Bazin, di Prony e di Eytelwein relative al movimento uniforme dell'acqua in un canale od in un fiume". *Atti della Società degli Ingegneri e degli Industriali* 4:203-214.

Russell, J.S. (1837). "Report of the committee on Waves". *Reports,* British Association for the Advancement of Science 7:417-496.

Russell, J.S. (1845). "Report on Waves". *Reports,* British Association for the Advancement of Science 14:311-390.

Saint-Venant, A. Barré de (1851). "Mémoire sur les formules nouvelles pour la solution des problèmes relatifs aux eaux courantes". *Annales des Mines* Série 4 20:183-357.

Saint-Venant, A. Barré de (1871). "Théorie du mouvement non permanent des eaux, avec application aux crues des rivières et à l'introduction des marées dans leur lit". *Comptes Rendus de l'Académie des Sciences,* Paris 73:147-154; 73:237-240.

Water Management and Hydraulic Structures in Antiquity: Contributions by Günther Garbrecht

Jürgen Garbrecht[1] and Henning Fahlbusch[2]

Abstract

Between 1967 and 1991, Dr. G. Garbrecht and collaborators investigated ancient hydraulic structures in the Eastern Mediterranean region. Four of these investigations are summarized herein. The summaries primarily focus on the description of hydraulic structures and the interpretation of their functionality and associated water management operations. Also, organizational aspects of the investigations and examples of the great value of the interdisciplinary approach to the discovery and interpretation of findings are presented. The investigations uncovered, even by today's standards, bold designs and planning of water resources systems, technical brilliance, grandiose architecture, and high-quality craftsmanship in the construction of the structures. It may be rightfully claimed that this work led by Dr. G. Garbrecht did not only increase our knowledge in the field of water resources development in antiquity, but also emphasized the absolute necessity of a real interdisciplinary approach in archaeological research.

Introduction

Born 1925 in northern Germany, Günther Garbrecht studied Civil Engineering at the Technical University Karlsruhe, Germany, where he obtained his Ph.D. in 1952. From 1954 until 1987 he was a member of the academic staff of the Technical University Istanbul, Turkey; the Middle East Technical University in Ankara, Turkey; the University of Zambia in Lusaka, Zambia; and, the Technical University Braunschweig, Germany. During his 15 years in Turkey (1954-1969), Dr. G. Garbrecht became fascinated by the impressive remains of hydraulic structures and water supply installations of ancient cities of the Near East. This fascination led him to engage in a second career, parallel to his main duties of teaching and research in hydraulics: the study of water-supply systems of ancient cities. After his retirement from academia in 1987, he continued to be active in the study of water supply systems and water management in antiquity. His most important work concentrated on the ancient city of Pergamon, Turkey. Additional excavations, surveys, reconnaissance and investigations

---

[1] USDA, ARS, Grazinglands Research Laboratory, 7207 West Cheyenne Str., El Reno Oklahoma, USA

[2] Fachhochschule Luebeck, University of Applied Sciences, Fachbereich Bauwesen, Stephensonstrasse 3, D-23562, Luebeck, Germany

were carried out in Egypt, Israel, Syria, Eastern Anatolia and Italy. Related field campaigns were conducted during the academic off-season and involved a mix of scientists, engineers, archaeologists and architects to allow for an interdisciplinary interpretation of the findings. Dr. G. Garbrecht secured grants from archaeological institutions and universities, and relied on personal contacts in the host countries to interact with authorities and obtain local assistance. From a research perspective, he attached great importance to an interdisciplinary treatment of the water supply problem by expanding on and including hydro-technological, archaeological and construction aspects in the studies. These efforts led to a better understanding of the planning, construction and operation of ancient water supply systems.

Dr. G. Garbrecht and his fellow scientists published numerous articles and books on ancient hydraulic structures and water management systems in the Eastern Mediterranean region. Cooperators in these studies were G. Holtorff, W. Brinker, H. Fahlbusch, K. Hecht, H. Thies, E. Netzer, H. Manderscheid, C. von Kaphengst, M. Döhler, H. Jaritz and A. Vogel. The results of his almost 30 years of investigations were also presented in post-graduate lectures (Technical University Braunschweig), extensive reports to the German Research Society (DFG), and the German Archaeological Institute (DAI), as well as in international symposia organized by him in France, Turkey, Greece, Israel and Egypt.

Fig. 1. Dr. G. Garbrecht, 1995.

The objective of this paper is to highlight selected contributions by Dr. G. Garbrecht to the history of hydraulic engineering. Findings and interpretations of his investigations in Pergamon, Urartu and Egypt (Sadd-el-Kafara and Fayum) are presented in this chapter. Projects relating to Jericho, Rome, Resafa and Dara are mentioned, but not covered in detail. References to these projects can be found in the cited literature. Due to the services he rendered in the field of the history of hydraulics, he has been awarded medals by the Istanbul Technical University (Turkey) and by the Frontinus Society (Germany).

The Water Supply of Pergamon

The investigation of the water supply of Pergamon was initiated in 1966, when Professor Dr. E. Boehringer, former President of the German Archaeological Institute (DAI), offered Dr. G. Garbrecht the opportunity to study the water-supply system of the historic city of Pergamon, located on the Aegean in western Anatolia, Turkey. During the two decades of research activities, an interdisciplinary working team was formed consisting of architects, historians of structures, surveyors, hydraulic engineers and students. This group of engineers and scientists was complemented by archaeologists of the DAI. Reconnaissance and surveys in the field were conducted during breaks in the academic year, generally in March/April or August/September. Important conditions for frictionless and successful activities in Turkey were the knowledge of the language, acquaintance with the local customs and, most important, reliable contacts with local authorities and institutions. These contacts were developed during Dr. G. Garbrecht's stay at the Middle East Technical University, Ankara, Turkey. The financial support of the research project was initiated by Dr. G. Garbrecht, and provided mostly by the German Research Foundation and by grants from the universities involved and other German research foundations. Financial support was awarded based on the scientific quality of the project, but was also influenced by the reputation and personal views of Dr. G. Garbrecht. In the following the discoveries and interpretations of the water supply system of Pergamon are presented.

Hellenistic Era. From 283-133 B.C., the city of Pergamon was the capital of the Hellenistic kingdom Pergamon. The fortified city was located on the south-western slopes of a steep, rocky hill, hereafter referred to as "city hill", located in the valley of the Kaikos River on the west coast of Anatolia, Turkey (Fig. 2 and for situation map see Fig. 3).

At about 200 B.C., the water demand of the royal city exceeded significantly the limited capacity of local water resources. To remedy the situation, three long-distance aqueducts were constructed north of the city, two originating in the Selinos Valley and one in the Madradağ Mountains. The aqueducts were 20 km to 45 km long, consisted of clay pipes with diameters of 13 to 18 cm, and had capacities of 6 to 45 liters per second (520 to 3900 m^3/day). The pipes were operated under free surface flow conditions. Figure 3 shows the general layout of aqueduct lines with trace #2 referring to two aqueducts from the Selinos Valley and trace #4 to the aqueduct from the Madradağ Mountains. Hydraulic calculations suggest that the net water delivery to the city hill of Pergamon from these three aqueducts may have been about 53 liters per second (4,600 m^3/day).

In order for the water to reach the top of the city hill, a 3 km wide depression that separates the Madradağ Mountain range in the north and the city hill (Fig. 2) had to be overcome by means of inverted siphons. The pipe material of the siphons was clay and stone for the two aqueducts originating in the Selinos Valley (trace #2 in Fig. 3), and lead for the aqueduct originating in the Madradağ Mountains (trace #4 in Fig. 3). Pressure heads of the inverted siphons were 25, 30 and 180 m, respectively. To the best knowledge of the investigators, this was the first time that the Archimedes principle of communicating pipes was applied in engineering practice.

Figure 2. The city hill of Pergamon (Acropolis), as seen from the Madradağ mountains north of the city. The remains of arcade bridge can be seen in the middle of the depression, and the inlet of the siphon is seen in the foreground.

The two aqueducts from the Selinos Valley (trace #2 in Fig. 3) were tracked, surveyed and investigated during the 1975/76 campaigns under the direction of Dr. G. Garbrecht. With regard to the Hellenistic Madradağ aqueduct (trace #4 in Fig. 3, and details in Fig. 4), only the existence and the general course of the aqueduct and the 3 km long siphon had been known since 1886. A survey, analysis and interpretation of the Madradağ aqueduct did not exist. Such an investigation was conducted between 1968 and 1972 by Dr. G. Garbrecht. The engineers of the team performed the surveys and hydraulic calculations, the architects provided a conceptual reconstruction of the bridges and other hydraulic structures, the archaeologists assisted in the dating of the structures,

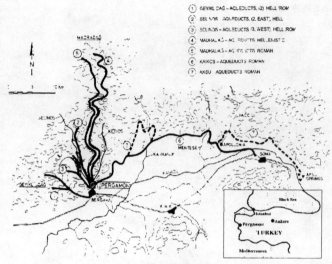

Figure 3. Long-distance aqueducts of the water-supply system of Pergamon.

and the philologists helped in the reading and interpretation of inscriptions and other written material.

The pipes of the siphon of the Madradağ aqueduct no longer exist today and were presumed to have been removed when the aqueduct ceased to operate. Also, the pipe material of the siphon was not known before 1972. Chemical tests of the soil beneath the siphon line were performed by chemical engineers and they established that the pipes were made of lead. The inverted siphon of Pergamon, with a pressure head of 180 m, is considered as a milestone in the history of hydraulic engineering. This siphon remained in operation during much of the Roman era.

Figure 4. The Hellenistic Madradağ aqueduct before reaching the depression.

Five additional aqueducts were discovered and surveyed in 1975/76 and 1981 by the team of Dr. G. Garbrecht. These aqueducts supplied water to the Asklepieion, a kind of medical and recreation center at the south-western edge of the city hill, and the surrounding area on the west bank of the Selinos River.

Roman era. In 133 B.C. Pergamon became part of the Roman Empire. During the long period of the "Pax Romana," Pergamon was no longer a fortress, but became a large open city. Population estimates range between 120,000 and 160,000 inhabitants. To meet the increased water demand, two more aqueducts were constructed, the Kaikos aqueduct which later became the Aksu aqueduct (trace #6/7 on Fig. 3) and the Roman Madradağ aqueduct (trace #5 on Fig. 3). The Kaikos and Aksu aqueducts were 53 and 73 km long, respectively, and had a capacity of about 150 liters per second (13000 m^3/day). The Madradağ aqueduct was about 50 km long and had a capacity of about 100-120 liters per second (9500 m^3/day).

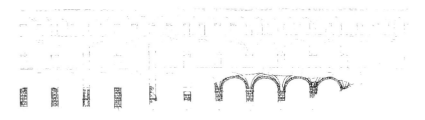

Figure 5. Conceptual reconstruction of the Roman arcade bridge across the depression north of the city hill. (Reconstructed and drawn by K. Hecht and H. Thies).

The Roman Madradağ aqueduct was discovered in 1969 by Dr. G. Garbrecht's team. The canals of all Roman aqueducts were constructed in masonry and plastered with mortar, and were situated slightly below ground surface. The Madradağ aqueduct was believed to have originated at a group of springs on the Toklu Mountain in the Madradağ Mountain range, about 5 km east of the Hellenistic aqueduct. About 10 km below the springs, the Hellenistic and Roman aqueducts ran parallel to one another until reaching the depression north of the city hill. To cross the depression, the Romans did not build a siphon, as the Greeks did 350 years earlier. Instead, the canal was led downhill, probably in the form of a cascade, and crossed the saddle at an elevation of 204.5 m MSL by means of a three-story arcade bridge as conceptually reconstructed by the architects of the investigating team (Fig. 5). In 178 A. D., only a few years after its completion, the arcade bridge was destroyed by a disastrous earthquake. It was reconstructed immediately and collapsed again during the earthquake of 262 A.D. It was then replaced by a quadruple pressure siphon and later (probably after 365 A.D.) by a double open channel on top of the remaining lowest arcade of the original three-story bridge. The dating and conceptual reconstruction of these phases of the arcade bridge was only possible through the combined input of engineers, architects and archaeologists.

Another aqueduct on the east side of Pergamon, the Kaikos Aqueduct in the Kaikos valley (trace #6 in Fig. 3 and Fig. 6), was known to exist. Dr. G. Garbrecht's team tracked and surveyed this aqueduct in 1973 and 1975. The Aksu aqueduct (trace #7 in Fig. 3 and Fig. 7) was discovered during this investigation in 1975. With regard to the Kaikos aqueduct, the slope was extremely small (on average 0.00034) and the course had to be kept as straight as possible. Therefore, the side valleys on the northern side of the Kaikos Valley were not followed, but crossed by means of bridges. Between Soma and Pergamon the remains of no less than 40 bridges have been found, the largest of them being the structure across the Karkasos River.

Figure 6. Kaikos aqueduct 13 km east of Pergamon.

Figure 7. Aksu aqueduct 16 km west of Aksu springs

The terrible earthquake of 178 A.D. damaged or destroyed most of the bridges of the Kaikos Aqueduct. In order to avoid the reconstruction of the large bridges and to reduce the damage by future earthquakes, the Kaikos aqueduct was redesigned with the following changes (trace #7, Fig. 3, Aksu Aqueduct):
- the upper part of the aqueduct was entirely relocated to the right bank of the Kaikos River;
- the flanks the Karkasos valley were followed with a smaller bridge higher up in the valley; and,
- the new stretches of the aqueduct were constructed as tunnels, several meters below ground surface (Fig. 7).

These changes led to a considerable reduction of the total number of bridges and, more importantly, eliminated the huge structures across the Kaikos River, the Yacili River and the Karkasos River (Fig. 2). The architects were able to conceptually reconstruct in detail five of the larger arcade bridges of the Kaikos/Aksu aqueducts.

Concluding considerations. The tracking, excavations, surveys and hydraulic calculations of the aqueducts by Dr. G. Garbrecht's team revealed an impressive example of the ingenious hydro-technical design and construction of a complex water supply system of an important city in the Eastern Mediterranean region. Development, climax and decline of the city extended over almost one millennium. The investigations of the Pergamon water supply system under the guidance of Dr. G. Garbrecht have demonstrated that a hydro-technical infrastructure was a vital component of successful ancient civilizations and should not be neglected in archaeological and historical research. The report describing the investigations and findings of the water supply system of Pergamon (Garbrecht, 2001) is considered a standard book on the water supply of ancient cities.

The Water Supply System of Ṭušpa (Urartu)

Many archaeological and historical studies have been conducted on the Kingdom of Urartu. An investigation that specifically targeted the water supply system of Ṭušpa was undertaken by Dr. G. Garbrecht alone during three visits in 1986, 1988 and 1991. The objectives were to complement previous studies by developing new insights based on water supply, water management and irrigation considerations. Each visit lasted only a few days and consisted primarily of touring various sites. Surveys or excavations were not conducted, and financial aid was not sought for this project. Contacts with local water authorities and permission for access were necessary and were greatly facilitated by a former student of Dr. G. Garbrecht who was native to the area. Based on these local site visits, Dr. G. Garbrecht examined the concept, planning and implementation of the sophisticated projects from a water management and engineering technology perspective (Garbrecht, 1987 and 1988). The primary contribution of Dr. G. Garbrecht to the existing literature was the understanding of the operation of the hydraulic structures and the water management of Lake Rusa. Part of this ancient water supply system is still in operation today.

Historical background. The earliest appearance of the name Urartu was in Assyrian inscriptions around 1250 B.C. The inscriptions identified the Kingdom of Urartu to be in the geographical area between the lakes Van, Sevan and Urmia in Eastern Anatolia, where the frontiers of Turkey, Iran and the former USSR meet (Fig. 8). As a result of continued attacks by the Assyrians from the south, the Hurrian tribes that lived in the Armenian highlands amalgamated around 850 B.C. to form the Kingdom of Urartu and rose to a leading position in the Near East. During the reigns of thirteen successive kings between 850 and 600 B.C., constant battles were fought against the great power of Assyria. Eventually Urartu fell in 595 B.C. following an assault by the Scythians and Medes.

Evidence from Assyrian inscriptions, annals and commemorative records of victories, as well as from Urartian inscriptions and archaeological excavations, indicated that the Urartian people were a nation with considerable artistic and technical skills. Special emphasis was repeatedly given to remarkable hydrotechnical installations in Urartu. The drinking and irrigation water supply of the Urartian capital, Ṭušpa, with water storage and long-distance transfer, was an outstanding example of thoughtful planning and excellent workmanship. The supply system was in operation for over 2,500 years and still serves nowadays, at least partly, its original purpose.

• URARTIAN SETTLEMENTS (TOWNS, FORTRESSES)

Figure 8. Approximate size of the Kingdom of Urartu (9th-6th century B.C.) in Eastern Anatolia.

The Menua Canal. Ṭušpa is situated in a plain on the eastern shore of Lake Van (Fig. 9). The water of the lake is undrinkable because of its high sodium carbonate content. The two creeks which flow into the plain of Van, Engusner Çayi and Kurubaş Çayi, do not carry water year round. In order to provide a secure and stable water supply, it was necessary to exploit resources farther away. The Uratian engineers first turned their attention to a large spring in the valley of the Engil Çayi, at a distance of about 54 km from Ṭušpa. Today the average spring flow is 6-8 m^3/sec, and does not fall below 5 m^3/sec, even during the dry season. In its natural course, the water from the spring flowed into the Engil Çayi, about 5 km away. The flow was captured - as it still is today - immediately below the spring by means of a simple stone and earthwork diversion, and was partly channeled over a bridge across the Engil Çayi to its right bank. Thereafter, the canal followed the course of the Engil Çayi along the northern flank of the valley in a westward direction. However, thanks to the many still existing old retaining walls (Fig. 10), it is possible to reconstruct the entire course of the old Urartian canal in complete detail.

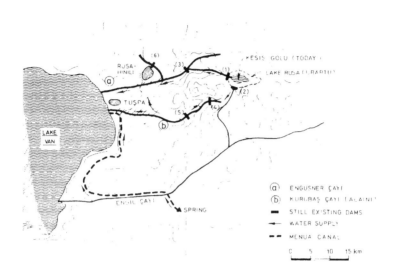

Figure 9. The plain of Van (hatched area). Water transfer to the two capitals Ṭušpa and Rusahinili via the Menua Canal and the two rivers Engusner and Kurubaş. Lake Van is located in the middle of the situation map on Fig. 8.

Figure 10. Retaining walls of the Menua canal, halfway between the spring and Ṭušpa.

About 25 km from the spring, the course of the canal turned north and followed the eastern shore of Lake Van (Fig. 9). Here the terrain presented a number of obstacles. Several narrow valleys had to be skirted, and rocky areas, as well as stone-swept slopes, had to be traversed. The size of the retaining walls of the canal along the valley flanks increased in proportion to these obstacles, and in certain cases was as high as 20 m. After a distance of about 54 km, the canal reached the plain and irrigated the gardens and fields of Ṭušpa, as it still does today. With an average canal capacity of 1.5 m^3/sec, a total volume of 45 $\circ 10^6$ m^3 could be conveyed annually to Ṭušpa. Because of the steady supply from the spring, it was not necessary to store the water. In its original shape and course and partly with its Urartian structures, the Menua Canal has been in operation for 2,500 years. It was only in 1950 that the middle segment of the canal was replaced by a modern concrete canal because maintenance of the old canal became too costly.

In general, it is difficult to date ancient water supply systems as they usually do not involve coins, pottery or other aids to help identify the chronology. However, the canal supplying Ṭušpa with water is an exception: there were fourteen inscriptions along the course of the aqueduct which state that King Menua (ca. 810-785/780 B.C.) constructed the canal. These inscriptions consisted of several lines of text, and were always found at locations where the construction was particularly difficult.

Water Storage in Lake Rusa. Around 700 B.C., the capital Ṭušpa was moved to Rusahinili (Fig. 9). However, the ancient citadel Ṭušpa, with its gardens and fields, was not abandoned, and the Menua Canal continued to supply the city with water. With regard to Rusahinili ("City of Rusa"), a new source of water was needed for the royal residence and its surroundings. Evidence of the date, concept and construction of the new water supply project was found on a gigantic stela discovered at the end of the 19th century in the surrounding hills of Lake Rusa (Fig. 9). The shallow lake is situated about 30 km away and 900 m higher than Rusahinili. According to the incomplete text of the stela, King Rusa (early 7th Century B.C.) had undertaken the following constructions:
♦ transformed a natural water basin into an artificial lake (Lake Rusa) by blocking the outflow at the natural outlet by a dam;
♦ directed the water from the lake to the new capital of Rusahinili;
♦ made the land around Rusahinili arable, and planted irrigated orchards and vegetable gardens;
♦ and, as a source of additional water, redirected water from the Alaini River (Kurubaş Çayi) to Rusahinili.

No traces were left of the original Northern Dam (#1 in Fig. 9) at the natural outlet of the lake. Based on the local topography, the dam, according to Dr. G. Garbrecht, must have been a large structure, 18 m high and 60 m long, with a wing dam 3 m high and 200 m long, extending over the right river bank. The main structure was probably earthfill with heavy rockfill protecting both exterior surfaces. The dam was repeatedly destroyed and rebuilt in the course of history. The last collapse happened in 1891. The dam was reconstructed in 1894/95 to a height of 3 m, and in 1952 to a height of 5.40 m (Fig. 11). The water released from the Northern Dam was directed into the Engusner Çayi which flowed directly to the garden town in front of Rusahinili hill.

Figure 11. The reconstructed dam at the northern outlet (to the Engusner Çayi) of the Rusa Lake.

The Southern Dam (#2 in Fig. 9) has survived to a great extent. It consisted of a core of earth, protected and stabilized by rockfill sections on the outside, each 7 m wide. The dam was about 5 m high, 27 m wide and 55 m long. No traces of flow-regulating devices were found at the outlet structure of the dam, which was an opening of 0.40 m width and 0.90 m height. The layout and elevation of the outlet structure indicated that the level of the lake in Urartian times was around 10 m higher than today. This implied that the crest elevation of the destroyed northern dam must have been considerably higher in Urartian times than it is today. The discharge from the Southern Dam normally reached Lake Van via the Engil Çayi. Dr. G. Garbrecht was of the opinion that the diverted water was redirected and transferred into the Kurubaş Çayi watershed for irrigation of the Ṭušpa-Rusahinili plain. He also believed that the Alaini River that was mentioned in the stela inscription was identical to the Kurubaş Çayi.

At least four more dams of medium size (# 3 to 6) were built in the two valleys on the east side of the Van Plain (Fig. 12), and about ten small-scale dams were reported to exist in the mountains between the Rusa Lake and the Van Plain. These were, according to Dr. G. Garbrecht, constructed at a later time (post-Urartian), probably after the first collapse of the Northern Dam of Lake Rusa. The average inflow into the Lake Rusa was estimated to amount to 20 million m^3 per year. Taking 40% losses (evaporation, seepage) into account, about 12 million m^3 water per year was available for irrigation and drinking for the cities of Ṭušpa and Rusahinili, and for the surrounding gardens and fields. A summarizing and concluding report by Dr. G. Garbrecht on the interpretation of the Urartian water management system is in preparation and will be published by the Deutsche Wasserhistorische Gesellschaft, Siegburg/Germany, in 2003.

Figure 12. Dam (#4 in Fig. 9) in the valley of the Kurubaş Çayi (post-Urartian).

Excavation and Survey of the Sadd-el-Kafara Dam in the Wadi Garawi (Egypt)

The existence of the remains of an ancient dam in the Wadi Garawi on the eastern bank of the Nile River near Heluan, some 30 km south of Cairo (see Fig. 16) has been known since the end of the 19th century. However, information on the dam consisted of only general descriptions and interpretations. During a visit in 1981, Dr. G. Garbrecht found that the remains of the dam had suffered greatly during the last century due to natural disintegration and human interference. To document the remaining parts of this unique monument, the structure was surveyed and investigated in the spring of 1982 by an interdisciplinary working group of nine German engineers and scientists under the direction of Dr. G. Garbrecht. Two former Egyptian students of Dr. G. Garbrecht established contacts with local authorities, obtained necessary permits, and provided language, custom and daily operational support required for a successful field campaign in a foreign country. Without their help this project would not have been possible. Request for financial support was initiated by Dr. G. Garbrecht and granted by the German Research Foundation. A detailed report on the excavation and survey of the Sadd-el-Kafara in the Wadi Garawi, and more general coverage of the dam, can be found in Garbrecht and Bertram (1983), and Garbrecht (1985).

The Dam. The dam was 113 m long and 14 m high, but today remains of the structure are only found on either side of the valley (Fig. 13 and 14). Only about 24 m of the northern wing of the dam extend into the wadi, and of the southern wing only about 27 m are preserved. Cooperating with the engineers was a team of scientists from the University of Applied Sciences, Karlsruhe, Germany, that conducted a detailed photogrammetric survey of the two wings. Between the remains of the wings a breach,

about 50 to 60 m wide, has formed as a result of numerous floods over the past 4,500 years. The surprising large width of 98 m may reflect the lack of knowledge and experience in dam building. It was probably believed that the width of the dam had to be related to the size of the lake or the volume of the water behind the structure. The cross-section of the Sadd-el-Kafara dam consisted of three elements, which differ in composition and function (Fig. 15):
- a central core of rubble, gravel and weathered material,
- two sections of rock fill on either side of the core, and
- layers of ashlars (0.30 x 0.45 x 0.80 m on an average) placed in steps on the slopes of the rock fill.

Figure 13. The Wadi Garawi with the remains of the Sadd-el-Kafara.

The outside face of the rock fill was without doubt one of the most remarkable construction features of the Sadd-el-Kafara dam. On the upstream side, the facing is still well preserved (Fig. 14). On the downstream side, isolated stone blocks indicated that a similar facing corresponding to that on the upstream side was planned and at least partially constructed. There were no traces of operational flow control devices, such as outlets or a spillway. If they existed at all, which is doubtful, they would have been placed in the missing center portion of the dam.

Assessments of the dam's stability by modern methods led to the conclusion that the design was basically correct, though very conservative. Construction of the dam was estimated to have taken about 10-12 years. This estimate took into consideration the available construction methods and technology, the volume of fill that had to be transported from the wadi terraces to the dam core, and the amount of rock that had to be brought from the nearby quarries to the slopes of the dam.

Figure 14. The Sadd-el-Kafara seen from the right bank of the Wadi Garawi. Flow direction from left to right.

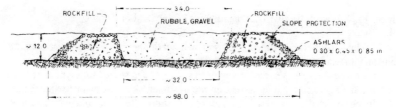

Figure 15. Cross-section of the Sadd-el-Kafara Dam.

Dating and purpose of the dam. Since the discovery of the Sadd-el-Kafara dam, there was no doubt that the dam was indeed a very old structure. Analyses of pottery and radiocarbon data obtained from samples of charcoal and textiles found in the remains of buildings northwest of the dam (probably the worker's camp during the construction period) indicated that the dam was built in the early Old Kingdom, about 2,700-2,600 B.C. This dating makes the Sadd-el-Kafara dam one of the oldest in the world. The careful investigation of the remains of the "Workers Camp", which led to the accurate dating of the dam, was conducted by archaeologists from the German Archaeological Institute and Swiss Archaeological Institute, both in Cairo, Egypt. The Jawa drinking water reservoir in Jordan may be older, and the diversion dams on the Kasakh River in

Southern Russia may be of an earlier date. However, the latter structures were much smaller, and the Sadd-el-Kafara may with good reason be called the world's oldest large-scale dam.

The purpose of the dam is still not quite clear. It may have been constructed for flood protection of a yet unknown site downstream at the mouth of the Wadi Garawi, or it may have been designed as a reservoir to support transport to the Nile Valley of alabaster ashlars which were quarried in the upper part of the valley.

Destruction of the dam. The 1982 investigations have uncovered evidence that the dam was never fully completed (Garbrecht, 1985). It seems that the structure was overtopped by a flood, at a time when the middle section of the downstream rock fill or even the core were still under construction. The collapse of the dam must have resulted in a catastrophic flood in the lower wadi. The impression left by the disaster, which was not caused by a faulty design but by a natural phenomenon that could not have been foreseen, must have led to the abandonment of the heavily damaged structure. This cooperative investigation lead by Dr. G. Garbrecht provided a more complete archaeological dating, conceptual reconstruction and engineering interpretation of the Sadd-el-Kafara Dam.

Water Storage for Irrigation in the Fayum Depression

The Fayum Depression west of the Nile Valley covers an area of about 1800 km^2 (Fig. 16). It is located 75 km south of Cairo and is still supplied today with water from the Nile River by the Bahr Yusef (Joseph's Canal). Surplus water that is not used in the settlements of the Fayum or for irrigation flows today, as in the past, into the low lying evaporation lake Birket el-Qarun situated in the north west of the Fayum depression. In addition, a reservoir, the El Mala'a reservoir, was established in the Fayum which allowed Nile water to be temporarily stored (Fig. 17). Also, the existence of a lake, called Lake Moeris, was reported by Herodotus (about 484-425 B.C.), Diodorus Siculus (1st century B.C.), Strabo (about 63 B.C.-26 A.D.), Pomponius Mela (1st century A.D.), and Pliny the Elder (24-79 A.D.) There are few ancient installations which are as controversially discussed as the Lake Moeris. Not only are the location and size of this lake in dispute, but even its existence is questioned. Yet, Lake Moeris may well be the El Mala'a reservoir.

Figure 16. Satellite photo of the lower Nile River and the Fayum Depression.

The controversy surrounding this question and his fascination for historical hydraulic structures induced Dr. G. Garbrecht to explore the remaining hydraulic structures in the Fayum and contribute to the interpretation of the associated water management operations. Investigations of the remaining structures were initiated by Dr. G. Garbrecht and performed in spring 1988 by the Leichtweiss-Institute for Water Research of the Braunschweig Technical University (Germany) and the Swiss Archaeological Institute, Cairo (Egypt). The research activities were supported by the Water Research Center, Cairo, the Ain Shams University, Cairo, and the University of Applied Sciences, Lübeck (Germany). Egyptian colleagues and former students of Dr. G. Garbrecht provided critical support in dealing with local authorities and institutions. The details of the investigations and results can be found in Garbrecht and Jaritz (1990, 1992) and Garbrecht (1996).

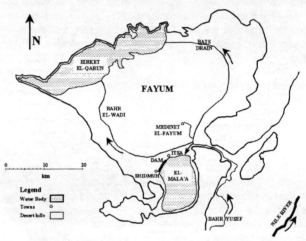

Figure 17. The ancient El-Mala'a reservoir situated in the south-east of the Fayum Depression.

The dam in the south-east of the Fayum. Available historical and archaeological information, as well as onsite inspections of the still existing sections of the dam, allowed interpretation of claims by both Diodorus and Strabo regarding the existence of a manmade reservoir for irrigation water storage with flow into and out of the reservoir controlled by gates. The Birket el-Qarun could not have been that lake because it did not have features suitable for water storage management. Also, in antiquity large-scale irrigation was possible only by gravity, and the Birket el-Qarun is located in the lowest part of the Fayum depression. Therefore, the higher-laying eastern parts of the depression were the only suitable sites to store water for irrigation purposes. From a purely topographic point of view, only the south-east portion of the Fayum present opportunities to establish and operate a storage basin. A wide basin that is separated from the Fayum depression by a low drainage divide exists at that location. Walls or dams about 8 km long and 7.5 m in height would be sufficient to create a reservoir with a capacity of 250 to 300 million m^3 (Fig. 17).

Results of the technical and archaeological surveys of the existing dam sections were:

- about 4 km of the 8 km long dam remained today. The wings of the dam, assumed to be only 1 to 2 m high, have not been found. It was reasonable to believe that they were not masonry walls but earth embankments.
- Plaster on the upstream face (applied to obtain water tightness), buttresses on the downstream side (arranged to ensure stability) and the general form of the cross-section of the wall (trapezoidal with vertical upstream face and stepped downstream face) clearly indicate that the wall served to retain water. Sediment deposition 1 to 2 m high on the upstream side of the wall was evidence that a lake had existed for quite a long time.
- A distinction was made between retaining walls strong enough to resist water pressure, and earth-fill structures protected by facing walls. Retaining masonry walls, with foundations on bedrock, were found mainly where water depths exceeded 3 to 4 m. Earth structures protected by facing walls, with foundations 0.7 m deep on soil, were preferred for dam heights less than 3 m.
- In at least three but probably four places, the dam failed several times. The extensive scours below 10 m wide breaches are proof of uncontrolled outflow of large quantities of water (Fig. 18).
- The most northern rupture is located where the wall was about 7 m high. Remains of five walls erected in succession in the same place (the last one is still intact to almost its full height) showed that in this place the dam was destroyed and rebuilt at least four times.
- The middle and the southern breach were each closed by massive brick walls with narrowly spaced buttresses. A 116 m repair of the southern wall is still in very good condition today (Fig. 19). The reconstructed sections indicate that the highest water level of the lake likely reached 16.5 m or 17.0 m MSL. Based on the 16.5 m elevation, the lake surface was 114 km^2 and the storage volume was 275 million m^3.
- At two locations in the region of the middle breach, structures were found that might have been water outlets and served as culverts after the lake had been abandoned. The northern culvert had three openings each 3 m wide and the southern one had three openings each 1.5 m wide. Both structures together had the capacity to completely evacuate the storage volume of 275 million m^3 during the four spring months of the second irrigation season.

Figure 18. View from the west over the water-filled scour (about 100 m wide and 250 m long) downstream of the northern breach. In the background is the still existing wall, and in the water are blocks of older damaged walls.

Dating of the dam. Lacking stratigraphic data, which is most helpful in archaeological excavations, an attempt was made to determine the date of construction by comparison of the walls with masonry work of other structures for which the age was reliably known. Such a study of the walls at the northern breach led to the following construction sequence:

Destruction of Wall No. (1)	3rd Century A.D.
Re-construction of wall No. (2)	6th Century A.D.
Destruction of wall No. (2)	702/704 flood
Re-construction of wall No. (3)	after 704 A.D.
Destruction of wall No. (3)	1159 flood
Re-construction of wall No. (4)	after 1245 A.D.
Destruction of Wall No. (4)	2nd half of 18th Century
Re-construction of Wall No. (5)	1st half of 19th Century

This dating applies to the five massive dams founded on rock at the northern breach. Over long sections, however, there also exist earthfill dams with up to three protective walls at the upstream face. It was not possible to directly determine the time of construction of the earthfill dams. Only comparative considerations were used to approximate the date to be before the period of the Roman masonry structures. Finally, the question of whether the first earthfill dam was built in the reign of Ptolemy I (305-283 B.C.) or Ptolemy II (285-246 B.C.), or whether it was constructed as early as the 26th dynasty (664-525 B.C.) during which the country enjoyed a period of great economic prosperity, remains unanswered. The timeline and dating of the five walls of the dam in the south-east of the Fayum, their association with the Nile floods and the general historical development of the Fayum in Ptolemic and Roman times was performed by H. Jaritz from the Swiss Archaeological Institute in Cairo, Egypt.

Figure 19. Repair wall with buttresses closing the breach.

Management of the El-Mala'a reservoir. With 275 million m^3 of stored water from the El-Mala'a reservoir and an assumed irrigation efficiency of 50 percent, a net area of about 150 km^2 could be irrigated for a second harvest in spring. For Roman times, a total irrigated area in the Fayum of about 1200 to 1300 km^2 is generally assumed. Thus, with the water stored in the El-Mala'a reservoir (the largest basin within the Fayum), 15 percent of the total farmland in the Fayum could be irrigated a second time.

It should be emphasized that the reservoir in the El-Mala'a basin was the largest but not the only reservoir in the Fayum region. As far back as 2000 years ago, the Fayum was a region in which water for two or more harvests per year was available. This was an optimal use of the limited water and soil resources, and was again achieved in the Nile valley only after the construction of the Aswân Dam in 1971. The El-Mala'a basin was not a permanent lake, as was the Birket el-Qarun, but a temporary reservoir. The reservoir was filled from August to December, stored water until February/March and was emptied again from March to May. During the period May to August, the reservoir bottom was certainly used for agricultural purposes.

The temporary water storage in the El-Mala's reservoir explained an apparent contradiction with regards to a second "lake made by man" in the Fayum, the so-called "Lake Moeris." Travelers did or did not see the lake, depending on the season during which they visited the Fayum. Cartographers did or did not include it on their maps, depending on whether they considered it to be a lake or a huge irrigation reservoir. The same applies to descriptions by local and foreign historians and geographers.

El-Mala'a - the legendary Lake Moeris? The question about the legendary Lake Moeris can best be answered if the elevation of the water level of the Birket el-Qarun evaporation lake between 1800 B.C. (12th dynasty, Amenemhet III, King Moeris) and the time of the Ptolemaic construction activities is taken into consideration. The following well-founded interpretation may, with justification, be assumed to be correct:
- from the 12th dynasty until Ptolemy I, the water level in the Birket el-Qarun was, on average, 18 m MSL, and absorbed part of the Nile flood;
- around 300-250 B.C. the water level was lowered to about 2 m below sea level with a corresponding increase in agricultural area. The lowering was achieved by restricting the amount of water entering the Fayum depression from the Nile;
- in Roman times water level of the lake was further dropped to about 36 m below sea level.

If these interpretations are correct, then Herodotus, who visited Egypt in 445 B.C., appears to have seen the state of the Fayum as it existed under the Pharaohs. All other ancient writers describe the Fayum after the Ptolemaic development measures. Diodorus Siculus was in Egypt in 60-56 B.C.; Strabo was in Egypt in 25/24 B.C.; Pomponius Mela (geographical compensium "De chronographica libri tres") was in Egypt around 43/44 A.D.; and, Pliny compiled part of his encyclopaedical "Naturalis historia" from books and collected the rest on his travels to Egypt between 50 and 70 A.D. The descriptions from Diodorus to Pliny do not leave any doubt that in the last century B.C. and in the 1st century A.D., a man-made lake, which temporarily stored flood water from the Nile for irrigation purposes, existed in the Fayum. The description of the lake, the canal and the mode of operation all apply to a temporary lake in the El-Mala'a area of the Fayum which dates back to Ptolemy I and II.

Thus, field investigations into the water-retaining structure between Itsa and Shidmuh led us back to the pre-Roman (Ptolemaic) epoch. The findings were in agreement with the reports of the classical writers. Hence, the question posed at the start of this section can be positively answered: there are no apparent reasons that preclude the identification of the lake in the El-Mala'a basin to be the lake the ancient (post-Ptolemaic) historians and geographers referred to as Lake Moeris. The successful interpretation of the Fayum water management scheme and the answer to the existence of Lake Moeris was primarily made possible by the close and harmonic cooperation between engineers, archaeologists and architects of Dr. G. Garbrecht's team.

Other investigations

A number of additional investigations on water-supply systems of ancient cities and on hydraulic structures were conducted by Dr. G. Garbrecht and collaborators, but are not presented within the framework of this paper. Examples include: the water supply of the Royal Winter Palaces and a large estate near Jericho, Palestine, about 150 B.C. to 50 A.D. consisting of three long-distance aqueducts and associated water distribution systems (Garbrecht and Peleg, 1994; Garbrecht and Netzer, 1991a and 1991b; Garbrecht, 1989; and, Netzer, 1999). The layout and operation of the water system of the Caracalla Baths in Rome, probably about 217-537 A.D. (Garbrecht and Manderscheid, 1994; and, Manderscheid and Garbrecht, 1991). The remarks on the crescent dam of Dara in southern Anatolia, Turkey (Garbrecht and Vogel, 1992). And, the water supply system of Resafa-Sergiupolis in Syria, 4th to 9th century A.D. (Brinker, 1991; Garbrecht, 1991; and, Brinker and Garbrecht, 2003 (in print)).

Summary

Between 1967 and 1991, Dr. G. Garbrecht, together with a group of fellow scientists and engineers, investigated eight ancient water resource development projects in the Eastern Mediterranean region. Analysis and evaluation of the research results demonstrated the ability of the technicians in those times to plan, design and execute outstanding hydraulic structures to ensure a safe and reliable water supply for human consumption and irrigation. Even today, after more than 2000 years, laymen and experts admire the boldness of the designs and planning of water supply systems, as well as the technical brilliance, grandiose architecture, and high-quality craftsmanship of the construction and execution of the necessary hydraulic structures. This assessment of the ancient water utilization projects is even more admirable, if one considers the limited engineering knowledge and the elementary construction methods of the times.

Until recently, engineering was, in comparison to the study of the arts, architecture, history, philology and epigraphy, considered only as a discipline of secondary importance. Pipes, canals, cisterns, dams and the like were merely registered but not examined and valued at the same level as the contemporary remains of architecture and the arts. It was not realized that the historical engineers established the indispensable water resources infrastructure that the great civilizations of the past required if they were to develop and exist. Only in the last three to four decades has this attitude towards engineering started to change, and technical scientists were increasingly accepted as indispensable members of archaeological research teams. It may be rightfully claimed that the research work of Dr. G. Garbrecht and his cooperators, as summarized in this chapter, not only increased our knowledge in the field of water resources development in antiquity, but also emphasized the absolute necessity of a real interdisciplinary approach in archaeological research.

References

Brinker, W. (1991) *Zur Wasserversorgung von Resafa-Sergiupolis.* Deutsches Archäologisches Institut, Damaszener Mitteilungen, Band 5, p. 119-146, Verlag Philipp von Zabern, Mainz.

Brinker, W., and G. Garbrecht. (2003) *Die Zisternen-Wasserversorgung von Resafa-Sergiupolis.* Accepted for publication by Deutsche Wasserhistorische Gesellschaft, Siegburg/Germany.

Diodorus, Siculus. (1814) *The Historical Library of Diodorus, the Sicilian.* Translated by G. Broth, Vol. 1, London.

Garbrecht, G. (1985) "Sadd-el-Kafara: the world's oldest large dam". *Water Power and Dam Construction,* Vol. 37, July 1985, Business Press International, Sutton England, (ISSN 0306-400X).

Garbrecht, G. (1987) "Die Talsperren der Urartäer". *In: Historische Talsperren,* Bd. 1, p. 139-145, Verlag Konrad Wittwer, Stuttgart (ISBN 3-87919-145-X).

Garbrecht, G. (1988) "Water/Management for Irrigation in Antiquity (Urartu, 850 to 600 B.C.)". *Irrigation and Drainage Systems,* No. 2, p. 185-198, Kluwer Academic Publishers, The Netherlands.

Garbrecht, G. (1989) "Die Wasserversorgung geschichtlicher Wüstenfestungen am Jordantal". *Antike Welt,* Heft 2, Mainz.

Garbrecht, G. (1991) "Der Staudamm von Resafa-Sergiupolus". *In: Historische Talsperren* Bd. 2, p. 237-248, Verlag Konrad Wittwer, Stuttgart (ISBN 3-87919-145-X).

Garbrecht, G. (1996) "Historical Water Storage for Irrigation in the Fayum Depression (Egypt)". *Irrigation and Drainage Systems,* No. 10, p. 47-76, Kluwer Academic Publishers, The Netherlands.

Garbrecht, G. (2001) *Die Wasserversorgung von Pergamon.* Vol. 2, Deutsches Archäologisches Institut, AvP I.4, Verlag Walter de Gruyter, Berlin/New York (ISBN 3-13-016947-9).

Garbrecht, G., and Bertram, U. (1983) *Der Sadd-el-Kafara, die älteste Talsperre der Welt (2600 v.Chr.).* Mitteilungen des Leichtweiss-Instituts für Wasserbau der Technischen Universität Braunschweig, Heft 81 (2 Vol.), Braunschweig (ISSN 0343-1223).

Garbrecht, G., and Jaritz, H. (1990) *Untersuchung antiker Anlagen zur Wasserspeicherung im Fayum/ Ägypten.* Leichtweiss-Institut für Wasserbau der Technischen Universität Braunschweig, Mitt.Heft 107, Braunschweig (ISSN 0343-1223).

Garbrecht, G., and Jaritz, H. (1992) "Neue Ergebnisse zu altägyptischen Wasserbauten im Fayum". *Antike Welt,* Vol. 23, Nr. 4, p. 238-254.

Garbrecht, G., and Manderscheid, H. (1994) *Die Wasserbewirtschaftung römischer Thermen (Archäologische und hydrotechnische Untersuchungen)*. Leichtweiss-Institut für Wasserbau der Technischen Universität Braunschweig, Mitt.Heft Nr. 118, Vol. A, B, and C, Braunschweig (ISSN 0343-1223).

Garbrecht, G., and Netzer, E. (1991a) "Die Wasserversorgung des geschichtlichen Jericho und seiner Winterpaläste". *Bautechnik*, Vol. 68 (1991), Nr. 6, p. 183-193, Berlin.

Garbrecht, G. and Netzer, E. (1991b) *Die Wasserversorgung des geschichtlichen Jericho und seiner königlichen Anlagen (Gut, Winterpaläste)*. Leichtweiss-Institut für Wasserbau der Technischen Universität Braunschweig, Mitt.Heft Nr. 115, Braunschweig (ISSN 0343-1223).

Garbrecht, G., and Peleg, J. (1994) "The water supply of the fortresses in the Jordan Valley". *Biblical Archaeologist*, Vol. 57, Nr. 3, p. 161-170.

Garbrecht, G. and Vogel, A. (1991) "Die Staumauern von Dara". In: *Historische Talsperren*, Bd. 2, p. 263-276, Verlag Konrad Wittwer, Stuttgart (ISBN 3-87919-145-X).

Herodot. (1984) *Neun Bücher der Geschichte. Klassiker der Geschichtsschreibung*, Phaidon Verlag, Essen (ISBN 3-88851-035-X).

Manderscheid, H., and Garbrecht, G. (1992) "Etiam Fonto Novo Antoniano". L'acquedotto Antoniano alle Terme di Caracalla. *Archaeologia Classica*, Vol. XLIV, L'ERMA Bretschneider, Roma.

Netzer, E. (1999) *Die Pläste der Hasmonäer und Herodes' des Großen*. Verlag Philipp von Zabern, Mainz (ISBN 3-8053-2011-6).

Pliny. (1854) *Naturalis Historia*. Published by Teubner, Leipzig (translated by Bostrock and Riley, London, 1855).

Pomponius Mela. (1646) *De situ orbis*. Lugd. Batav, V.9 and XXXVI.16.

Strabon. (1829/1835) *Geographie*. Publisher Metzler, Stuttgart (translated by K. Kärcher, Stuttgart).

Hans Albert Einstein's Efforts to Understand and Formulate Bed-Sediment Transport in Rivers

Robert Ettema and Cornelia F. Mutel[1]

Abstract

Hans Albert Einstein (1904-1973) might have remained one of countless civil engineers whose work, although locally important, had little impact on the world as a whole. His career as a structural engineer in Germany in the late 1920s started in that manner. However, his trenchant independence of spirit and famous father, Albert Einstein, pulled him from this position and launched him into a productive career as researcher and educator fascinated with the mechanics of bed-sediment movement and water flow in rivers. Hans Albert Einstein made several milestone contributions, including the formulation of the first comprehensive method relating rates of bed-sediment movement and water flow.

Introduction

By virtue of the times in which he lived, the trans-Atlantic span of his life, and his name, the career of Hans Albert Einstein's (hereinafter called Einstein) forms a convenient course along which to chart the advance of alluvial-river mechanics as an engineering science during the twentieth century. The present paper follows part of that course by focusing on Einstein's efforts to understand and formulate a central issue in alluvial-river mechanics – the relationship between bed-sediment discharge and water discharge.

Einstein's career was strongly marked by historic movements; each career change was induced by a change in the political climate linked with such movements. He left Germany (1931) and Switzerland (1938) in large part because of apprehension about growing Nazi power. His father, Albert, initiated both moves, understanding well the threat posed to Jews in Europe. Einstein, whose undergraduate education was as a civil engineer, began his professional interest in alluvial river mechanics just when Europe was plunging into the Great Depression, a period when jobs for civil engineers severely declined, and when the Swiss were feeling especially pressed to address problems with the Alpine reach of the Rhine. His arrival in the U. S. coincided with the recent establishment of the Soil Conservation Service (SCS), reflecting a great national concern about soil erosion and the condition of many

---

[1] Both IIHR - Hydroscience and Engineering, College of Engineering, The University of Iowa, Iowa City, IA, robert-ettema@uiowa.edu, connie-mutel@uiowa.edu

rivers. That concern was set aside momentarily during World War II, but returned urgently right after the war, at which time Einstein was well positioned to play a leading role in addressing it.

Family correspondence reveals that although Albert first dissuaded his son from entering civil engineering, he later fostered and directed that career. Albert helped Einstein locate jobs, as a structural engineer in Germany (1927), a graduate research assistant in Switzerland (1931), and a research engineer in the U. S. (1938). The last two positions channeled Einstein into alluvial-river engineering. Subsequently, during the early 1940s, Albert encouraged his son to become a protégé of Theodore von Kármán at the California Institute of Technology, though for several reasons that did not eventuate.

Within the decade following his arrival in the U. S. (c.1938 – 1948), Einstein emerged as a leading expert in alluvial-river mechanics, his expertise being widely sought. This paper explains in part how that came about. For technical assessment of Einstein's overall contributions to alluvial-river mechanics, the writers defer to Shen's (1975) useful synopsis, and to practically every major textbook on alluvial-river mechanics. A symposium proceedings, held to honor him on the occasion of his retirement, lists his publications and the graduate students with whom he worked (Shen 1972).

Beginnings – Meyer-Peter's Flume

Professor Eugene Meyer-Peter of the Swiss Federal Institute of Technology (ETH) in Zurich needed to know how much sediment moved with water flow along the Alpine Rhine, especially the amount of coarser sediment, gravels and sands that along the river's bed. He intended to modify the river. In the late 1920s, the Swiss federal government and the local cantonal government of St. Gallen, responding to concerns about an alarming increase in the frequency with which the river flooded, had contracted him to recommend an effective modification to the Alpine Rhine over a 20-kilometer reach extending from the Alps to the head of Lake Constance.

The river as it wends through the Swiss Alps to Lake Constance (Fig. 1) is the principal tributary to the Rhine, which flows on to Germany and The Netherlands. Bathymetric surveys to determine the elevations of the river's bed had revealed that the bed and therefore the river were rising, reducing the cross-sectional area of river channel available for water flow. The surveys were showing that the river was depositing substantial portions of its incoming sediment load. Evidently, the amount of sediment entering the river reach was more than its flow of water could move. By depositing its sediment, the river was aggrading, steepening its slope to enable its water to convey more sediment. This behavior, however, could not be tolerated. Extensive river control works over prior years had straightened and confined the river between levees, which could not accommodate an elevated river whose bed already was higher than some parts of its floodplain. Left unchecked, the aggrading river quite likely would break out of its leveed banks, disastrously flood the valley, and possibly form a new channel off to the side of its present course.

Political considerations gave Meyer-Peter's work urgency, as the Alpine Rhine above Lake Constance forms an international border between Switzerland, Austria, and Liechtenstein. An unregulated shift in the channel would open up

Figure 1. The Alpine Rhine constrained to a single, straightened channel just upstream of Lake Constance.

awkward negotiations about border location and land exchange. The negotiations between Switzerland and Austria had been complicated enough in 1900 when a bend in the river was straightened to reduce flooding at a small Swiss town about 10 kilometers upstream from Lake Constance. The tensioned political climate of 1930 was not conducive to such negotiations.

Besides continual dredging, the only practical way to deepen a loose-bed river is to narrow it and let the flow make its own bed. The key issue weighing on Meyer-Peter's mind, therefore, was how much to narrow the channel. Tackling that issue would require a comprehensive investigation of water flow and bed-sediment transport, because in 1930 there existed no reliable design equations for sizing alluvial river channel so that they would convey their loads of water and sediment. Prior efforts at similar channel modification in Europe had been hit or miss, essentially relying on intuition or overly simple rules-of-thumb to select channel width. The problems with the Alpine Rhine clearly showed that rules-of-thumb and "feel" were not sufficiently dependable. More understanding of fundamental processes was needed.

Similar efforts had been underway elsewhere, notably by the British who sought an improved design method for alluvial, irrigation canals cut through sandy terrain in parts of the Indian sub-continent and Egypt. The effort, termed the regime method (e.g., Lacey 1929), relied almost entirely on empirical relationships to characterize channels under long-term equilibrium or "regime."

The regime method was still in development, and its applicability to the Alpine Rhine with its gravel bed remained uncertain.

Meyer-Peter implemented a comprehensive plan of field measurements, hydraulic modeling, and flume experiments to be conducted in ETH's new hydraulics lab. To recruit research assistants, he placed an advertisement in a Zurich newspaper. The ad caught Mileva Einstein's attention, and she contacted her former husband, Albert. He had briefly thought and written about aspects of river mechanics (Einstein 1926), appreciated the importance of Meyer-Peter's work, and saw a promising, safer career opportunity for his and Mileva's elder son. In 1931 Einstein joined Meyer-Peter's research effort.

Though at first a rather lackadaisical and playful research assistant, not that well regarded by Meyer-Peter, Einstein eventually became intrigued by gravel movement along Meyer-Peter's flume (Fig. 2). After a few years, Einstein and colleagues wrote a series of papers presenting research findings stemming from Meyer-Peter's plan. The papers were on gravel transport, hydraulic radius and flow resistance, measurement of sediment transport, and hydraulic modeling (Meyer-Peter et al. 1934, Einstein 1934, Einstein, 1935, Einstein 1936, Einstein and Muller 1939).

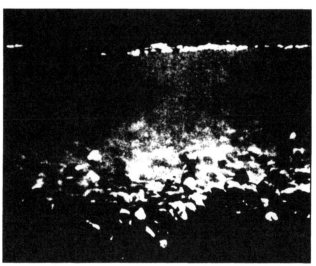

Figure 2. An observation, by Einstein, of 22-mm-diameter gravel from the Alpine Rhine moving in Meyer-Peter's new flume at ETH.

Einstein attained the doctorate degree from ETH in 1937 (Einstein 1937), submitting a novel thesis demonstrating the relevance of probability theory for describing the movement and transport of bed sediment, notably for relating sediment transport rates to the step motions of individual bed particles and flow turbulence. Though the scope of his thesis research did not include formulation of a method for estimating rates of bed sediment transport under given water flow and channel conditions, Einstein's insights into individual motion of bed particles formed the basis for his new approach to formulating the component of bed-sediment transport commonly called bedload transport, the transport of bed particles in successive contact with the bed. Under sufficiently intense flow conditions, beyond those for his ETH studies, bed particles may travel substantial distances suspended by flow turbulence. Such suspended load and bedload together comprise the total bed-sediment load conveyed by water flow.

Formulation – Enoree-River Flume

Albert, who had moved to the United States in 1934 because of concern for political movements in Germany, persuaded Einstein to come to the United States in 1938. Albert assisted his son find a position as a co-operative agent with SCS's newly established field laboratory on the Enoree River (Fig. 3), near Greensville, South Carolina. The lab was established for measuring sediment loads in the Enoree River in order to better understand the relationships between sediment transport and water flow. It was located in a region of South Carolina that had experienced severe soil erosion problems incurred with intensive farming. Einstein worked with colleagues Joe Johnson and Alvin Anderson on ways to measure sediment transport. They quickly realized the need to distinguish two populations of sediment conveyed by water in the river. In a paper (Einstein et al. 1940) they coined the term "washload" to describe the river's load of suspended fine silt and clay-size particles derived from soil erosion and usually not comprising the river's bed, the source of bed-sediment load.

Einstein continued converting the findings from his thesis research into a method for describing and predicting bedload transport of sediment in rivers and streams. He was convinced that the critical-shear-stress approach used by several prior formulations (e.g., Shields 1936) was physically inadequate. In his opinion bedload movement was better related to flow turbulence near the bed. Accordingly he took the principal conclusions of his thesis work and used them as a basis for a radically new formulation approach that equated the volumetric rate of bedload transport to the total number and volume of particles in motion. In turn, the number and volume of particles in motion depend on the probability that water flow would lift or eject individual particles from their seating on the bed and move them downstream in a given period.

The difficulty lay in determining the probability that the hydrodynamic lift on any particle on the bed is about to exceed the particle's weight within a given period of time. From a different perspective, the probability could be viewed as the part of the bed for which hydrodynamic lift force exceeds particle weight. The probability problem comprised two parts. One part concerned the need for an equation for hydrodynamic lift; particle weight is relatively easy to formulate. The other part concerned finding a meaningful expression of time; transport rate implies movement per unit time. Einstein adapted a well-known and standard formula for hydrodynamic lift, writing it in terms of a local velocity of water flow at a level near the bed. Here, though, assumptions were needed regarding estimation of the velocity and lift coefficient.

Figure 3. SCS's Enoree-River Flume, South Carolina, 1939. The flume was designed for measuring sediment loads and flow in an actual river.

A trickier problem concerned the inclusion of a time period. The most reasonable period to use is the average time required for the water to remove one particle from the bed. Unfortunately there is no way to express the time required for hydrodynamic lift to pick up a particle. Einstein assumed that lift involves some characteristic dynamic feature of the flow field around a particle falling in still water. Particle diameter divided by particle fall velocity expresses a characteristic time. Up to this stage his formulation was reasonably rigorous, once the under-girding assumptions about average particle step length were accepted. But the subjective use of fall velocity for particles in a description of particles rolling and bouncing along the bed was unsettling.

By combining formulas for the volume of bed particles moving as bedload, hydrodynamic lift on a bed particle, bed particle weight, and characteristic time based on bed particle fall velocity, Einstein arrived at formulas indicating the existence of a relationship between two parameters characterizing bedload transport. The two parameters, ϕ (phi) and ψ (psi), were central to Einstein's characterization of bedload transport; ϕ is a dimensionless expression for intensity of sediment discharge, and ψ is a dimensionless expression for flow intensity or gross shear force exerted on the bed. Einstein could not derive the exact form of the relationship between ϕ and ψ. Too many factors were unknown. Instead he had to find the relationship from plots of bedload data interpreted as ϕ versus ψ. If his formulation were correct conceptually, the data would lie systematically along a single curve signifying a single general relationship, or "law," for bedload transport. It then would be a fairly straightforward to fit an equation to the curve, thus expressing the relationship mathematically.

A Validation Test – Gilbert's Flume

Obtaining reliable data from which to determine the relationship, however, was not straightforward. Einstein used the only two comprehensive sets of lab flume data readily available to him at the time: his own from ETH, and those published by Karl Grove Gilbert over twenty-five years earlier (Gilbert 1914). Gilbert, a protégé of John Wesley Powell, had conducted novel and comprehensive flume experiments at the University of California – Berkeley. The great river surveys of the 1800s (notably Humphreys and Abbot 1861, Powell 1875) were accompanied by engineering and scientific desire to know more about the mechanics of rivers in the U. S. Gilbert's data encompassed a greater range of sediment and flow conditions than did Einstein's ETH data for gravel transport.

By and large the two sets of data fell along a curve in accordance with his formulation, except for a range of conditions reflecting high intensities of sand transport. Those data veered substantially away from Einstein's postulated curve, and clustered along their own curve. The deviant data disconcertingly suggested that bedload transport could not be described using his method. What also disconcerted him was the realization that the deviant data were not merely a batch of results from a set of extreme hydraulic conditions, but in fact were representative of flow and sediment transport in the sand-bed channels representative of most rivers in the U. S.

The deviation caused Einstein to review the formulation of his method, and to question the accuracy of Gilbert's data for sand bed channels. He queried his own assumption that all bedload particles moved in steps of constant length proportional to particle diameter, unaffected by flow conditions. His work at ETH had suggested this to be the case for the gravel beds at fairly low intensities of transport for which the probability of particle entrainment was moderate or low. He conjectured that, with increasing intensity of transport, the probability of entrainment is high and the step lengths increase from the constant length at low intensities. As step length increases, the area and number of particles starting movement together increases, and consequently so does the rate of bedload transport. This refinement of his theory modified the relationship between ϕ and ψ, and led to a second curve with a common stem as the original curve, but which veered away in almost the same manner as the cluster of Gilbert's sand-bed data. The new curve, though, still did not run through those data. Einstein wondered if Gilbert's data were tainted with measurement error. More experimentation and field data were needed.

Further Calibration – Mountain Creek

In contrast with the Alpine Rhine and the Enoree, Mountain Creek is a mere ditch (Fig. 4). Yet, to Einstein, Mountain Creek was an ideal little river. The ways in which the creek conveyed its water and sediment, its response to changing patterns of land use in its watershed, and its response to seasonal fluctuations in weather mimic those of large rivers. Therefore the creek possessed most of the characteristics of alluvial channels that Einstein sought to understand and formulate. Moreover, it was conveniently small so that Einstein could measure its water and sediment loads. The Enoree River field station had proven disappointing for obtaining field data on bedload because of insufficient large flows

Figure 4. Mountain Creek, Mississippi, 1941. The creek was fitted with a size-reduced version of the sediment-measurement apparatus used for the Enoree River.

Mountain Creek was a useful stepping-stone to understanding and formulating flow and sediment transport in large rivers, such as the Alpine Rhine or the Missouri. He needed the creek if he was to advance his ideas about sediment movement in rivers. The creek could help in calibrating or linking his laboratory-flume insights and equations to the behavior of a full-scale sand-bed river. He had learned from his Alpine-Rhine work at ETH that laboratory results, and formulations based only on the results of laboratory idealizations of rivers, usually are regarded skeptically by practical engineers dealing with real rivers. If he could show that his ideas worked for sediment movement in Mountain Creek as well as in his ETH flume, then showing that they worked for a river would be a matter of simple geometry.

As perverse luck would have it, the summer and autumn of 1941 were relatively dry in South Carolina. Flows in the creek barely moved any sediment. Only a few inches of rain fell, though a single storm did drop an inch-and-a-half of rain during the evening of almost the last day Einstein intended to monitor the creek. He and technicians were out at the enlivened creek immediately the next morning, recording its discharge of rainwater and sediment. The equipment worked well and the measurements proved, at least to Einstein's satisfaction, that his concepts seemed to be valid for a small stream like Mountain Creek (Einstein 1944). He was heartened additionally by subsequent measurements taken from West Goose Creek in Mississippi.

By 1942, Einstein had sufficiently ordered his thoughts on a method for describing and predicting bedload transport that he was able to get them published as a paper (Einstein 1942). As was the practice of the ASCE *Transactions Journal* at the time, appended to his paper were discussions by researchers interested in alluvial sediment transport. His paper drew praise for its attempt to relate sediment movement and flow mechanics, but it raised questions about the main assumptions leaping the gap between formulation concepts and presentation of a practical predictive method. In particular, the paper drew a salvo of comments for purporting to be on greater rational basis than were the current formulations. Difficult questions included – Why base the formulation on lift force alone? Why should settling velocity be included in a formulation of bedload transport? One discussor (Kalinske) remarked that Einstein evidently had "stepped over into the realm of abstract dimensional analysis" when he uses particle settling velocity as a convenient parameter to put the probability of particle motion in a time context. Kalinske (1942) also had attempted to include turbulence in formulating bedload movement of sediment.

Method Extended – Caltech Flume

With the entry of the U. S. into World War II, and after the modest yield of results from the Enoree River, SCS wound down its Enoree Field Station and re-assigned its personnel. In 1943 Einstein went to Pasadena to SCS's collaborative laboratory at Caltech. Albert was enthused about the move and encouraged his son to contact von Kármán, a renowned Caltech fluid mechanician. Von Kármán, however, was busy with war-related matters.

As his part of the war effort, Einstein was seconded to Caltech's Hydrodynamics Laboratory to work on shock waves produced by explosives and projectiles breaking the sound barrier. However, he still had opportunities to continue developing his bedload method and to investigate several pressing problems emerging in the wake of dam building and other engineering activities along rivers, in particular along the Rio Grande River. As Einstein saw things, accurate estimation of bed-sediment load was the most important problem in alluvial-bed river engineering. The ability to predict alluvial sediment movement in a river would enable engineers to predict the river's response to changes in its water and sediment loads, thereby reducing the uncertainty associated with utilizing the river as a resource for water and hydropower.

Convinced of the essential correctness of his bedload method, Einstein set about extending it by addressing several complicated aspects of bed-sediment transport – bedform development, transport of nonuniform bed sediment, and combined bedload and suspended-load transport of bed sediment.

A fellow SCS researcher at Caltech, Hunter Rouse, earlier had formulated and published an SCS report with an equation accurately describing the vertical distribution of suspended bed sediment over the depth of flow (Rouse, 1939). The equation, now often called the Rouse equation, is one of the more successful formulations of sediment transport. However it gives only the distribution of suspended-sediment concentration relative to some reference elevation near the bed, showing that the concentration decreases rapidly with higher elevation in a flow. Concentration of sediment usually is orders of magnitude less at the water surface than near the bed. The difficulty from a practical standpoint was that the equation does not give the absolute suspended-sediment load. To get that, the relative distribution has to be referenced to a known, or estimated, sediment load concentration near the bed. Here, Einstein saw an opportunity to link the so-called Rouse equation for suspended sediment load with his formulation of bedload and, in his words, to produce "a unified method for calculating the part of the sediment load in an alluvial river that is responsible for maintaining the channel in equilibrium" (Einstein 1950). The formulation approach used for his first attempt (Einstein 1942) would have to be adjusted. Yet more insights and data were needed.

Einstein surmised that the suspended-load distribution as described by Rouse's equation could be spliced to the top of the bedload layer as described by the bedload formulation. The concentration of particles moving at the top of the bedload layer would serve as a convenient and reasonable reference concentration with which to set the maximum concentration at the bottom of the suspended load distribution. However, to splice bedload and suspended load to form a unified method for estimating total load of non-uniform bed sediment meant that Einstein needed to put his bedload model on a somewhat more analytical basis. The statistical laws about particle motion he observed at ETH would have to be modified and restated, including his earlier assumption – the average step of a certain particle seems always to be the same even if the hydraulic conditions or the composition of the bed changes. Gilbert's data and the data from Mountain Creek and West Goose Stream showed that this assumption did not hold at the intense rates of sediment transport occurring for sand-bed rivers and streams.

Einstein recognized that he needed more flume experiments. Further field work would be useful to the extent that field data are always useful for developing and verifying theory, but it would not shed the necessary light on those fundamental aspects of sediment transport physics that need to be understood in order to make progress. Flume experiments were needed especially to see how particles move under conditions of intense rates of sediment transport, and to get an adequately detailed set of data on bed-sediment transport at high intensity rates for which bed sediment would be transported in suspension as well as along the bed. Gilbert's study contained data taken for high intensity rates, but his data were not accompanied by all the measurements necessary to process the data in terms of Einstein's unified method. During the period 1944 through 1946, water and sediment were churned through a recirculating flume in the SCS lab for a series of experiments with mixtures of sand, to produce data on intense rates of sediment transport.

Recognition – Rio Grande River

In May 1947, the various federal agencies concerned with rivers and their watersheds convened at the Denver headquarters of the U.S. Bureau of Reclamation (USBR) to hold the nation's first meeting focused on the sedimentation troubles facing engineers and soil conservationists in the U. S. All the federal agencies sent representatives. Also in attendance were engineers and scientists from diverse state agencies, universities, and a number of overseas organizations. The conference placed Einstein center-stage as one of the nation's few leading authorities on sediment transport at a time when the full implications of the nation's sediment troubles were becoming pressing.

The need for the conference had arisen from a growing national recognition of the widespread, adverse consequences that sediment troubles were posing for river-basin development and for the conservation of land and water resources. The national scope of the troubles had become increasingly worrisome during the mid-1930s, shortly after the federal government had initiated numerous programs to enhance irrigation, hydropower, navigation, flood control and soil conservation. Severely eroded watersheds, river-channel aggradation or degradation, reservoir sedimentation, and the adverse environmental effects of muddied waters all indicated that much more needed to be learned about watershed and river behavior. Particularly prominent among the sediment troubles discussed were those engaging Einstein along the Rio Grande River.

SCS sedimentation colleague Vito Vanoni (1948) outlined for conference participants a history of the development of predictive relationships for sediment-transport and water flow in alluvial rivers. He laid out the big questions to be addressed in order to better understand how rivers move sediment, and ended his presentation by lamenting the lack of scientific and engineering attention given in the U.S. to sediment-transport problems, problems whose national importance he ranked with the more popular contemporary problems of atomic energy and rocket propulsion. Less than ten professionals in the U. S., he estimated, were devoting the major part of their time to the study of sediment transport. Prominent among the ten was Einstein, whose new approach Vanoni described as "a radical departure from all previous bed-load formulas."

Einstein addressed the participants on two issues of keen interest with regard to the sediment troubles of the Rio Grande: measuring and predicting the rate at which rivers move sediment along their bed (Einstein 1948). Accurate measurements were needed in order to know how much sediment a river moves under existing flow conditions. Prediction is important when knowledge is needed of sediment movement under altered flow conditions, such as when a river is to be narrowed. He likened the middle Rio Grande to Mountain Creek and West Goose Creek, asserting that the Rio Grande behaves essentially like the two creeks. Unlike the other speakers, Einstein could draw on and describe European experience as well as experience in the U. S. Moreover, for many participants the name Einstein held beguiling promise of major breakthroughs in understanding and formulating the mechanical laws of sediment transport by rivers. Within several months of the Interagency Sedimentation Conference, former SCS colleague Joe Johnson facilitated an opportunity for Einstein to join the engineering faculty of UC-Berkeley.

Over the following two years, Einstein completed for SCS a detailed write-up of his bed-sediment transport method and published it as U. S. Department of Agriculture Report 1026 (Einstein 1950), now widely recognized as a milestone in alluvial-river mechanics. His method became widely known thereafter as the Einstein method, used extensively by the USBR, the U. S. Corps of Engineers (USACE), the U.S. Geological Survey (USGS), and others. It implemented several substantial advances beyond the bedload method in Einstein (1942). Report 1026 not only presented an elegant splicing together of bedload and suspended-load components of bed-sediment transport, but also introduced new concepts that reduce some of the empiricism in the 1942, ϕ-versus-ψ relationship. The concepts include – modifying flow intensity parameter ψ to ψ', the latter parameter involves only an estimated portion of the flow's energy expended on bed-particle roughness (not on the entire bed); introduction of adjustment factors to account for local velocity, and pressure distributions at a bed of non-uniform sized sediment; and, estimation of transport rates for size fractions of sediment.

Application – Missouri River

Though the Einstein method laid out in Report 1026 was comprehensive, the method was somewhat cumbersome to apply, and some of its components were found to need adjustment. Einstein discovered this when working with USACE engineers to apply it to the Missouri River. For two months during the summer of 1948, about four months after his first meeting as a board member for the Missouri River's Sedimentation Studies, Einstein was in North Dakota and Montana, at the headquarters of the USACE's Garrison District. Over the remainder of his career he maintained a productive relationship with USACE, assisting it with sediment concerns along the Missouri, the Arkansas, and other rivers.

USACE's Missouri River Division was charged to oversee a vast watershed, stretching, at one end, from the confluence of the Missouri and Mississippi Rivers to the Rocky Mountains at the other end. The Missouri's watershed covers about one fifth of the continental U. S. In 1948, the division's efforts were concentrated largely on implementing the Pick-Sloan Plan to control and regulate the river's flow for the purposes of flood control, hydropower generation, and navigation. However, the division found this process to be fraught with more difficulties than the plan had envisioned. All of the dam projects called for under Pick-Sloan were facing difficulties and setbacks attributable to the river's sediment. Consequently, the division's mandate to implement the Pick-Sloan plan was beleaguered with unresolved questions about how to cope with the sediment.

By early August 1948, the technical problems facing the division were clear to Einstein. He summarized them in a brief report to the division, stating that "every attempt must be made to study the questions in the Missouri itself and in its tributaries. This is the only way to find the relative importance of the various influences" (in Board minutes). He realized in particular that the veracity of the predictive methods would have to be checked by the simultaneous measurement of sediment transport and the flow variables of the natural river.

Figure 5. Board Members (Einstein, Straub, Vanoni, Lane) and Corps engineers ponder the Missouri River at its release from Ft. Peck Dam, Montana, 1948.

Einstein proposed that the river's sediment be lumped into two distinct populations, bed-sediment load and washload, each to be treated separately. The populations originate from different sources, and are subject to different sets of mechanical relationships between water flow and sediment transport. Washload could be readily estimated from measured concentrations of fine sediment in the river's water column. Bed-sediment load, however, posed a long-standing measurement problem.

He also proposed options for measuring bed-sediment load in a big river like the Missouri. One option would be to measure the amount of bed sediment conveyed as suspended load in the river. He knew that a substantial part of the river's sediment load moves in suspension, and that there exists no major difficulty to measure suspended load down to about half a foot from the bed. The experiments he conducted while with the SCS had shown that if he could measure the suspended load, he could extrapolate an estimate of the bedload, thereby making it feasible to estimate bed-sediment load all the way to the bed.

A second option for measuring bed-sediment load would be by successively surveying the growth of the delta region at the head of reservoirs along the river. A river's bed sediments typically deposit along a delta, whereas the washload sediments, being fine, usually spread out over a wide area of the reservoir, often by means of density currents. A delta deposit, therefore, approximately represents the aggregate of bed-sediment load flowing into the reservoir. If the hydraulic conditions of the river were known for the period of sediment deposition in the reservoir, it would be possible to verify the accuracy of the method used for determining bed-sediment loads. An accurate method would help in assessing the river's response to the dams and other engineering works. A study to check this option soon was underway for the Niobrara River, a tributary of the Missouri.

One immediate matter was a little delicate. The method selected for estimating the rates of bed-sediment transport through the river was the bedload equation proposed by Professor Lorenz Straub, a prominent hydraulics engineer who had been a USACE engineer in the 1930s. Straub (1935) had proposed the method while working on House Document 238 (Missouri River Report), a detailed assessment of the flow and sediment problems posed to engineering use of the Missouri River. As Straub was the senior, and initially the most vocal, board member, and since his method had been developed expressly with the Missouri River in mind, his was the method that the division had decided to adopt. Einstein was uncomfortable with the method. In a long letter report to the division, he outlined the steps that needed to be taken to gauge the sediment load conveyed by the river, and he went through the shortcomings of Straub's method. Besides being essentially an extension of the old, shear-stress excess approach proposed in 1879 by the French engineer Paul Du Boys, Straub's method assumed that the river kept its cross-sectional shape and its roughness for the full range of water flow and while the river's bed degraded or aggraded. These assumptions seemed unreasonable to Einstein.

The idea that the river's bed would be of constant roughness was not supported by initial calculations of flow depth and flow rate for selected reaches of the river. Those calculations showed that the Manning's resistance coefficient, n, for the bed reduced systematically from about 0.03 at the lowest flow rate to some value near 0.015 for the highest flow rate. In physical terms, using a constant value of 0.03 substantially retarded the river's capacity to convey water flow at the higher discharge rates, and did not match reality. Einstein together with the division's engineers had examined data on the Missouri River at Fort Randall, South Dakota, as well as data from its tributaries the Niobrara, Big Sioux, and Elkhorn Rivers in an effort to find a possible common relationship explaining the variation between the roughness value (Manning's n) for these rivers. A first attempt to explain it on the basis of changing channel shape failed, because the channel shape did not change appreciably. A further attempt related flow energy loss to the intensity of bed-sediment transport and thereby Einstein's modified flow-intensity parameter ψ'. This attempt proved more promising, but needed lab-flume proof.

Both Straub's and Einstein's methods eventually were used for estimating bed-sediment transport, though Straub's was soon abandoned. Einstein, though, encountered an unexpected complication: downstream of Ft Peck Dam, the river's degrading bed became armored with coarser sediment. No bed-sediment method had taken armoring into account.

Further Refinement – Berkeley's Flumes

UC-Berkeley's strong reputation and its capacity to attract talented graduate students, combined with his link to USACE, enabled Einstein to undertake at Berkeley a sustained research effort aimed at better understanding and formulating sediment-transport processes. It was an effort largely undertaken by graduate students and USACE engineers working under Einstein's guidance. They embarked on a comprehensive series of flume investigations aimed at illuminating key aspects of sediment and flow behavior. Moreover, Einstein's Berkeley appointment enabled him to teach, something he enjoyed (Fig. 6).

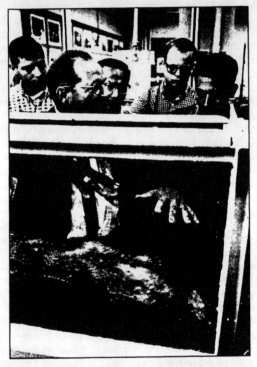

Fig. 6. Einstein at a Berkeley flume explaining flow processes to students.

Briefly mentioned here are two examples illustrative of that effort. An especially important issue, and one that has challenged formulation of sediment transport in rivers, concerns what happens if the bed sediment comprises a wide range of particle sizes. This situation, of course, is the norm for most riverbeds. A basic assumption underpinning the Einstein method needed further work; i.e., that all particle sizes in a river may be equally available at the bed surface and within the bed. Over about one year, 1950-1951, graduate student Ning Chien and Einstein carried out a series of flume experiments, and were in a position to provide detailed descriptions of the processes whereby different sized bed particles segregate in the upper layer of a river bed, how armoring occurs, and how a riverbed acts much like a reservoir for sediment, storing it during periods of reduced water flow, and releasing it during periods of greater water flow.

With doctoral student Robert Banks, Einstein began investigating how several factors contribute to flow resistance in rivers. This effort would provide more insight for his sediment-transport method, and it would help address a crucial companion issue concerning the relationship between water discharge and depth in alluvial rivers. The total resistance opposing the flow consists of the combined effect of resistance attributable to surface roughness, bedforms (or bar resistance as he expressed it), and vegetation. Einstein wondered if the total resistance could be expressed as the sum of these components. This thought was not new. It had been used successfully in determining flow resistance in flow around bodies. To investigate how different resistance elements combine to resist flow, Banks and Einstein undertook a series of flume tests in which they placed different forms of roughness. Their tests showed that the total resistance exerted by the combined types of roughness equaled the sum of the resistance forces.

The notion of dividing flow resistance into two parts, particle roughness drag and bedform drag, was new for alluvial-river mechanics. Its first practical implementation is the Einstein-Barbarossa method for estimating the relationship between flow depth and flow rate in alluvial channels. Einstein and Nicholas Barbarossa, an engineer with the USACE's Omaha District, used field data from the Missouri, several of its tributaries, and two California rivers to find a relationship between Einstein's parameter ψ' and that part of flow-energy loss attributable to bedforms. They used the Manning-Strickler equation to estimate energy loss attributable to surface roughness. Publication of their method (Einstein and Barbarossa 1952) is an acknowledged milestone in alluvial-river mechanics.

Confronting Complexity

Einstein's effort to understand and formulate the relationship between rates of bed-sediment movement and water flow represents an historic departure from the largely empirical approaches that prevailed at the time. He introduced into the relationship the emerging fluid-mechanic concepts of turbulence and boundary layers, and of statistics, concepts that were becoming well established in engineering during the early decades of the twentieth century. He was not alone in this effort. Through the 1930s (e.g., Hunter Rouse, Anton Kalinske, Albert Shields), and especially after World War II, an increasing number of people had implemented advances in fluid mechanics in efforts to understand and formulate aspects of alluvial-river mechanics. Their efforts produced numerous improved insights and formulations of component mechanisms, and improved methods for estimating river behavior. However, at times their efforts were received skeptically.

The complex mix of mechanisms at play in natural alluvial rivers has defied (so far at least) reliable prediction of bed-sediment transport and flow depth; uncertainties of 100% or more are common for predictions of bed-sediment load. Engineers and scientists have long recognized that rivers are complex, and accordingly have used largely empirical as well as analytical approaches to characterize alluvial-river behavior. Commonly, the practical design engineer and the scientist in the field have found the empirical approach more useful and have been skeptical of sophisticated, predictive methods based on advanced fluid mechanics and data from laboratory flumes. Proponents of the more empirical approach and those of the largely mechanistic approach are quick to debate each other's methods, especially when one claims to be the superior. The following exchange is an example of the debate, and illustrates Einstein's conviction about the ultimate truth of the concepts supporting his method for estimating bed-sediment load. The exchange follows a paper published by Ning Chien, Einstein's student.

Shortly before he returned to China where he was to play a leading role addressing that country's river problems, Chien (1954) published two papers that that drew a salvo of criticism from a leading exponent of the (empirical) regime method approach to river behavior. One paper (*The present status of research on sediment transport*) addressed the issue of relationship between water discharge and bed-sediment load. Chien essentially argued that to answer the question, it is necessary first to be able to formulate flow in an alluvial channel, then to formulate how the flow affects the sediment comprising the channel's bed. Appended to Chien's ASCE *Transactions* paper is stern discussion criticizing his neglect of the body of understanding collectively termed the regime method. The discusser, Thomas Blench, a very capable hydraulic engineer and a leading proponent of that method, argued that Chien's paper presented knowledge limited only to findings from "laboratory flumes with trifling flows." He further argued that Chien had neglected "the vast amount of observations on canals in the field, the dynamical aspect of the formulas evolved there-from, and the fact that these formulas provide a simple and adequate means of practical design that has been used widely for many years." The regime equations, Blench claimed, "represent what real channels actually do."

Einstein, though not a coauthor of Chien's paper, wrote an additional closure discussion to that by Chien. He took issue with the Blench's claim about the adequacy of the "superiority of the 'simple and adequate'" regime formulas. Those formulas, he pointed out, were developed by curve-fitting of data from "a very narrow range of bedload conditions." He went on to express, among other things, his doubt that the regime formulas would work for rivers in the U. S. In his closure following Blench's discussion of another paper (Einstein and Chien 1954), Einstein presented figures showing how inadequately the regime formulas performed.

Since Einstein (1950), numerous methods have been developed for estimating the relationships between water discharge and bed-sediment transport in alluvial rivers. Some methods have built on the method laid out in Report 1026, or modified the method for better accuracy and more convenient use (e.g.,

Colby and Hembree 1955, Bishop et al. 1965). Others have developed from improved insights, and still others have remained resignedly empirical (e.g. Brownlie 1982). Ironically, Meyer-Peter's research plan for the Alpine Rhine led directly to a method for estimating bedload transport (Meyer-Peter and Muller 1948). Their method, essentially empirical, is simpler to use and practically as accurate as the Einstein method when applied to estimate lower intensities of sediment transport. It does not, however, enable estimation of suspended load.

Figure 7. Einstein maintained a life-long interest in sediment movement.

Hans Albert Einstein was not the mythical scientist his father became, and he was by no means the only engineer or scientist to have sought to understand and formulate river behavior. Much of his career is the archetypal story of the researcher protagonist determined to master intellectually the way water flows and conveys alluvial sediment in a river. In that effort, he personifies the mixed success and frustrations experienced by many researchers who have attempted to describe the complicated behavior of alluvial rivers in terms of rationally based equations. The effort begins enthusiastically enough, with apparent good promise of success, based on new insights into component processes. Formulation seems within reach, and significant progress indeed is made, but it soon involves compromises with empiricism. When formulation does not universally fit, the researcher devotes much of the remainder of his/her career tinkering with the formulation to get it to fit, or leaves thwarted. Throughout his career Einstein retained his fascination with alluvial rivers (Fig. 7) and his efforts to understand how they convey sediment. At the time of his death in 1973 he and friend Don Bondurant, a retired USACE engineer, had outlined a book on alluvial rivers. It was not to be the usual format of textbook, but rather an approach that introduces typical engineering problems arising between people and alluvial rivers, then explains the knowledge and methods needed to solve the problems.

References

Bishop, A. A., Simons, D. B. and Richardson, E. V. (1965). "Total bed-material transport." *Journal of Hydraulic Engineering, ASCE*, Vol. 91, HY2, 175-191.

Brownlie, W. R. (1981). *Prediction of flow depth and sediment discharge in open channels.* Rept No. KH-R-43A, W. M. Keck Laboratory of Hydraulics and Water Resources, California Institute of Technology, Pasadena.

Chien, N. (1956). "The present status of research on sediment transport." *Trans. ASCE*, Vol. 121, 833-868.

Colby, B. R. and Hembree, C. H. (1955). *Computations of total sediment discharge, Niobrara River near Cody, Nebraska.* U. S. Geological Survey, Water Supply Paper 1357.

Einstein, A. (1926). "Uber die ursachen der maanderbildung der flusse und Baer'sschen gesetzes." *Die Naturwissenschaffen*, 14, 223-225.

Einstein, H. A. (1934). "Der hydraulische oder profile-radius." *Schweizer Bauzietung*, Band 103, No. 8, 89-91.

Einstein, H. A. (1935). "Die eiching des im Rhein verwendeten gescheibefangers." *Schweizer Bauzietung*, Band 110, No. 12-15, 29-32.

Einstein, H. A. (1936). *"Der geschiebetrieb als wahrschienlichkeitsproblem."* Doctoral Thesis, ETH, Zurich..

Einstein, H. A. und Muller, R. (1939). "Uber die ahnlichkeit bei flussbaulichen modellversuchcen." *Schweizer Archive fur Angewandte Wissenschaft und Technik*, Heft 8, Zurich.

Einstein, H. A., Anderson, A. and Johnson, J. (1940). "A distinction between bed load and suspended load." *Trans. Amer. Geophysical Union*, 628-633.

Einstein, H. A. (1942). "Formulas for the transport of bed sediment." *Trans. ASCE*, Vol. 107, 561-597.

Einstein, H. A. (1944). *Bed-load transport in Mountain Creek.* U.S Soil Conservation Service, Tech. Paper SCS-TP-55.

Einstein, H. A. (1948). "Determination of rates of bed-load movement." *Proc. Federal Interagency Sedimentation Conference*, 75-90, Denver.

Einstein, H. A. (1950). *The bed-load function for sediment transportation in open channel flows.* U.S. Dept of Agriculture, Soil Conservation Service, Tech. Bulletin No. 1026, Washington, D.C.

Einstein, H. A. and Barbarossa, N. L. (1952). "River channel roughness." *Trans. ASCE*, Vol. 117, 1121-1146.

Einstein, H. A. and Chien, N. (1954). "Similarity of distorted models." *Proc. ASCE*, Vol. 80, 1-21.

Lacey, G. (1930). "Stable channels in alluvium." *Proc. Inst. of Civ. Eng.*, Part 1, 229, 259-292.

Humphreys A. A. and Abbot, H. L. (1861). *Upon the physics and hydraulics of the Mississippi River.* Pub. by J. B. Lippincott & Co., Philadelphia.

Gilbert, G. K. (1914). *Transportation of debris by running water.* Professional Paper No. 86, U. S. Geological Survey, Washington, D. C.

Kalinske, A. (1942). "Criteria for determining sand transportation by surface creep and saltation." *Trans. American Geophysical Union*, Part II, 266-279.

Meyer-Peter, E., Favre, H. and Einstein, H. A. (1934). "Neuere versuchsresultate uber den geschiebetrieb." *Schweizer Bauzietung*, Band 103, No. 4, 89-91.

Powell, J. W. (1875). *Exploration of the Colorado River of the West and its tributaries.* U. S. Government Printing Office, Washington, D. C.

Rouse, H. (1939). *An analysis of sediment transport in the light of turbulence.* U.S Soil Conservation Service, Tech. Paper SCS-TP-25.

Shen, S. W. (Ed.), (1972). *Sedimentation.* A Symposium to Honor Professor H. A. Einstein, Pub. H. W. Shen, Colorado State University, Ft Collins.

Shen, S. W. (1975). "Hans A. Einstein's contributions in sedimentation." *Journal of Hydraulic Engineering, ASCE*, 101(HY5), 469-488.

Shields, A. (1936). *Anwendung der ahnlichkeitsmechanik und der turbulenzforschung auf die geschiebebewegung.* Mitt. Der Preussischen Versuchanstalt fur Wasserbau und Schiffbau, Heft 26, Berlin.

Vanoni, V. (1948). "Development of the mechanics of sediment transportation." *Federal Interagency Sedimentation Conference*, 209-221, Denver.

A Tribute to Victor L. Streeter

D. C. Wiggert[1] and E. B. Wylie[2]

Abstract

Over the span of nearly half a century, Victor L. Streeter made significant contributions in fluid mechanics and applied hydraulics. He authored several popular textbooks and numerous technical publications. In this presentation we will first chronicle his career and then recount his efforts and lasting impact in the fields of hydraulics and fluid transients.

Victor L. Streeter

---

[1] Professor Emeritus, Department of Civil and Environmental Engineering, Michigan State University, East Lansing, MI 28824; Ph: 517-332-1790; Fax: 517-332-7346; wiggert@msu.edu

[2] Professor Emeritus, Department of Civil and Environmental Engineering, University of Michigan, Ann Arbor, MI 48109; ebw@engin.umich.edu

Introduction

Victor L. Streeter's professional career of 42 years included seven years of professional practice followed by 35 years in academia. His unusual skills in teaching, research, and writing provided a platform from which he could contribute in a lasting way to the technical understanding and professional practice of fluid mechanics and hydraulics. His ability to translate highly technical material into a form available to engineering practice through the authorship of three internationally recognized textbooks is probably his greatest legacy. These books, together with the editing of the *Handbook of Fluid Dynamics*, McGraw Hill (1961) have made his name rise to a level equal to the most prominent in our profession during the last fifty years.

He was born in Marcellus, Michigan November 21, 1909 where his early education was completed, followed by two years of college at Western Michigan University. In 1929 he transferred to the University of Michigan where he earned the BSE (CE) in 1931, MSE 1932, and ScD 1934. The Master's degree thesis was titled "Dimensional Analysis and Similitude", and the ScD "Frictional Resistance in Artificially Roughened Pipes". He and his wife Evelyn raised 2 children, Mary Reamer of Washington, DC and Victor J. Streeter of Ann Arbor, MI.

He was a registered Professional Engineer, an active member in the Professional Societies of ASEE and IAHR, a Fellow in ASCE and ASME, as well as a member of Sigma Xi, Iota Alpha, Phi Kappa Phi, Chi Epsilon, and Tau Beta Pi. His honors are numerous, and are listed in Table 1.

Table 1. List of honors and awards

1930 & 1931	Honors Convocation, University of Michigan
1935-36	ASME Freeman Traveling Scholarship, Göttingen and Karlsruhe
1936	ASCE Collingwood Prize for juniors (his doctoral thesis)
1952	Visiting Fulbright Lecturer, University of New Zealand, Christchurch
1953	University of Michigan Distinguished Alumni Citation
1966	James Clayton Fund Award, Institute of Mechanical Engineers, UK
1974	ASME Worcester Reed Warner Memorial Medal
1982	ASCE Hunter Rouse Hydraulic Engineering Lecture
1985	Victor L. Streeter Fellowship, University of Michigan

A Chronicle of the Career of Victor L. Streeter

Following his graduate studies at Michigan, he worked at the U. S. Bureau of Reclamation in Denver, initially on various theoretical and experimental studies in the hydraulic laboratory and later in an advisory capacity in connection with technical phases of hydraulic designs and operating problems. This covered the five-year span from 1934 to 1939, interrupted by an 18-month period, 1935-36, when he was the recipient of the John R. Freeman Travel Scholarship awarded by ASME. He first went to the Universität Göttingen, and studied fluid mechanics under Professor L. Prandtl of the Kaiser Wilhelm Institute for Flow Research. This was followed by a semester at the Karlsruhe Technische Hochschule with Professors Böss and Spannhake. During this period he visited many of the major hydraulics laboratories in Europe and the Orient. His translation of an article on artificial roughness studies by H. Schlichting was published in the *Proceedings*, ASCE, November 1937.

Prior to entering academia he gained two years additional experience in El Paso, Texas with the U. S. section of the International Boundary Commission, United States and Mexico. These involved river flows, flood protection, flood-water elevations and flood damage, as well as irrigation system designs including water conservation, planning and cost estimates for irrigation changes, structures, waste ways, etc. In September 1941, he changed his role from professional engineering practice, with the exception of consulting, and began an academic career that extended 35 years to his retirement in 1976.

His first position was Associate Professor of Hydraulics, Illinois Institute of Technology, Civil Engineering Department, where he was in charge of the hydraulic laboratory and taught courses in fluid mechanics, hydraulics, water supply, and mechanics. In 1945 he was appointed Professor of Civil Engineering and Mechanics. About the same time he became Chairman, Fluid Mechanics and Thermodynamics Division of Armour Research Foundation, responsible for all research projects in the division, as well as sponsor contacts and personnel relations. From 1947 to 1954 he was Research Professor and Director, Fundamental Fluids Research, IIT, and consultant in Mechanics to Armour Research Foundation. During this time, in addition to teaching, he was engaged in a variety of sponsored research projects, including turbulent flow, three-dimensional axially symmetrical flow, air dropping of supplies, wearibility of air force parachutes, and hydrodynamics of lubrication.

During this 14-year period he spent a brief period at the David Taylor Model Basin, Washington studying advanced fluid mechanics and developing the publication of his first well-known and well-referenced textbook, *Fluid Dynamics*, McGraw Hill (1948). This was followed closely with the publication of the first edition of *Fluid Mechanics*, McGraw Hill (1951) a book, with its subsequent editions, that became for an extended time period the most widely referenced textbook on fluid mechanics throughout the world. During 1952 he was the recipient of a Fulbright visiting lectureship to the Civil Engineering Department, University of New Zealand, Christchurch, N.Z., where he lectured on topics in advanced fluid mechanics, and contributed to laboratory development.

In 1954, he was appointed Professor of Hydraulics, Civil Engineering Department, University of Michigan, Ann Arbor, MI where he spent the remainder of his productive career, retiring in 1976. Initially, he continued teaching, writing, and doing research. However, in the mid 1950's when the digital computer was becoming a useful tool for engineering analysis he embraced its use with enthusiasm. He incorporated its use in courses and broadened his research to attack problems that were previously intractable. In the late 1950's he became interested in expanding his knowledge of water hammer through the use of the digital computer, and this became his primary focus for the rest of his career.

Steady Flow in Pipes and Conduits

Streeter's early scholarly work, including his ScD thesis and postdoctoral studies in Germany, initiated his lifelong interest in pipe flows. Those efforts dealt with steady flows in piping and conduits, focussing on means to more accurately and effectively account for the effects of frictional resistance, minor losses due to valves and fittings, etc. A treatise is provided in Chapter 6 of the volume *Engineering Hydraulics* (Rouse, 1950). Figure 1 shows a correlation between pipe friction factor and energy and momentum correction factors that he developed (Streeter, 1942).

Fluid Transients

Professor Streeter has made a lasting impact on the field of computational fluid transients. With E.B. Wylie as co-author, he has written three books on the topic: *Hydraulic Transients*, McGraw-Hill (1967); *Fluid Transients*, FEB Press, Ann Arbor (1982); and *Fluid Transients in Systems*, Prentice Hall (1993). A discussion of some of his significant contributions as well as his affiliations with colleagues and students follows. Rather than list these accomplishments chronologically, we chose to categorize them topically, keeping in mind that many of these subject matters were dealt with over extended periods of time, and indeed overlapped one another. The topics are not meant to be inclusive, but serve to illustrate some of the diverse subjects that intrigued and motivated him.

Computer Analysis of Water Hammer. An early publication in fluid transients dealt with computer analysis of water hammer in a pipeline due to rapid valve closure (Streeter and Lai, 1963). Other authors previously had published significant treatises on this subject, but this paper may have popularized the computer-based solution technique known as the method of characteristics (MOC) combined with specified time intervals for water hammer analysis. In that work Streeter collaborated with doctoral student C. Lai (1962, Appendix). Figure 2 shows predicted and experimental pressure waveforms measured in a pipeline with a dead-end branch. One defect in the numerical method discovered later was the attenuation of the predicted variables due to numerical interpolations. The thesis of D. Contractor (1963, Appendix) revealed the accuracy of non-interpolated predictions compared with experimental pressure waveforms measured in a laboratory pipeline, Figure 3. This and the former comparison demonstrates the accuracy of the numerical technique, and the manner in which line friction and minor losses can be included in the numerical model. Additional publications in particular dealt with analyzing pressure transients in pipe networks and distribution systems (Streeter, 1965-66, 1967), and use of mixed MOC/implicit methods (Streeter, 1969).

Turbomachinery and Water Hammer. Incorporating a centrifugal pump or turbine in water hammer predictions has been of prime importance for many industries. Of significance in a paper on this subject was the presentation of the complete dimensionless homologous characteristics for pumps with three specific speeds developed at the California Institute of Technology (Streeter, 1964). These curves

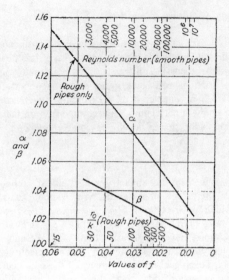

Figure 1. Energy and momentum correction factors for smooth and rough pipes (Streeter, 1942).

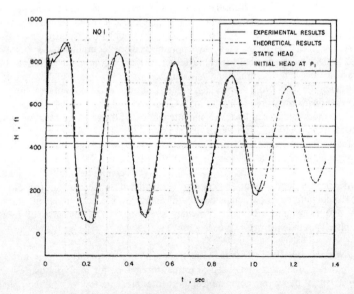

Figure 2. Experimental and numerical pressure wave forms (Streeter and Lai, 1963).

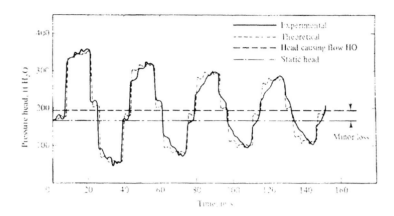

Figure 3. Experimental and numerical pressure wave forms (Contractor, 1965, Appendix).

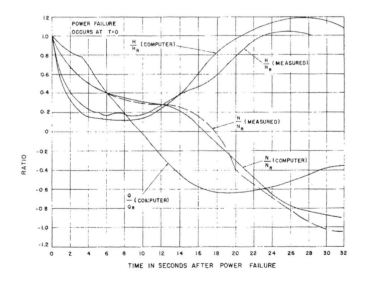

Figure 4. Tracy Pumping Plant failure (Streeter, 1964).

were incorporated into the water hammer analysis, along with the dynamics of the pump to predict power failure of one or more of a set of identical parallel pumps connected to a common suction pipeline and a common discharge line with arbitrary valving at the pump discharge. Figure 4 illustrates the well-known Tracy Pumping Plant failure predictions. In a later paper (Streeter and Wylie, 1975) the pump data was included in a computer algorithm using the Marchal, Flesch, and Suter transformation. Even though the interpretation of pump characteristics in that form was difficult, the transformation enabled a much simpler pump boundary algorithm to be formulated than the former method.

Valve Stroking. Usual water hammer analysis is performed by exciting a particular pipeline system and observing the attendant response. For example, the transient resulting from valve closure or system failure would be analyzed by imposing the valve motion or failure mechanism on the system. Streeter (1963, 1967a) pioneered an alternative approach, namely stroking the valve in a prescribed manner that permits certain constraints to be achieved. For instance, he demonstrated how a valve could be closed at the downstream end of a pipeline so that a maximum pipe stress is not exceeded and pressures remain above their steady-state values at all times. The technique allows for a piping system to be designed for arbitrary maximum pressure without back flow and without separation of the fluid column. The doctoral research of T. Propson (1970, Appendix) provided experimental data that validated the methodology for a wide variety of cases. Figure 5 from Propson's thesis illustrates the valve stroking technique.

Column Separation and Cavitation. A precise and accurate means to predict column separation in piping systems has eluded engineers for many years, even to this date. The phenomenon is complex, and may not be entirely deterministic in nature. Even though the rupture of a liquid column occurs near its vapor pressure, the presence of dissolved gases, impurities in the liquid, degree of liquid agitation, and incline of the piping all have a bearing on the outcome. Streeter ventured into this problem as a natural outcome of his comprehensive study of fluid transients. The detailed doctoral study of R. Baltzer (1967, Appendix) formed an experimental basis for some of the early work, and M. Weyler in his thesis (1969, Appendix) investigated additional losses attributed to bubble-like cavitation. Subsequently, working along with E.B. Wylie, Streeter's contributions include various investigations of the so-called classical single cavity model. In addition they developed a unique discrete vapor cavity model. The analytical version defines a rarefaction wave that causes vapor formation, and subsequently the vaporous flow is consolidated, that is compressed, by shock fronts of liquid moving into the vapor regions. The methodology is described in the Third ASCE Hunter Rouse Hydraulic Engineering Lecture (Streeter, 1982), and is illustrated in Figure 6.

Transmission Pipelines and Natural Gas Pipeline Transients. Along with his colleague E.B. Wylie and doctoral student M.A. Stoner (1968, Appendix), Streeter investigated transients in natural gas piping systems. In addition to employing the MOC, implicit methods were employed, either in combination with MOC or in stand-

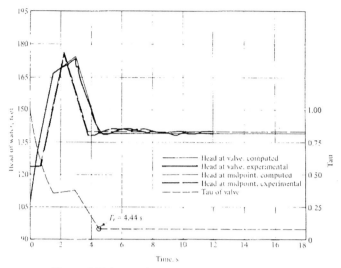

Figure 5. Valve stroking in a simple pipe system
(Propson, 1970, Appendix).

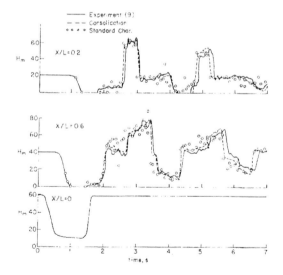

Figure 6. Pressure waveforms for column separation in a pipeline
(Steeter, 1982).

alone applications. Unique features included an integrated friction term (Streeter and Wylie, 1970) and the inertial multiplier introduced by W-S Yow in his thesis (1970, Appendix) that allowed the use of an increased time step in the MOC (Wylie, Streeter and Stoner, 1974).

Frequency Domain Analysis. Spearheaded by E.B. Wylie's doctoral research (1964, Appendix) Streeter contributed to several studies dealing with impedance methods and resonance in piping. Diverse systems were investigated, including a hydroelectric scheme (Wylie and Streeter, 1965), a liquid rocket engine configuration (Fashbaugh and Streeter, 1965), and oscillatory flow in piping that incorporated a reciprocating pump (Streeter and Wylie, 1966), see Figure 7.

Two- and Three-dimensional Transients. Another innovation introduced by Streeter was the algebraic MOC solution (Streeter, 1967b). It permits the characteristic equations to be applied over several reaches using the time increment of only one reach. One novel application was the analysis of low Mach number two- and three-dimensional transient flow using the one-dimensional algebraic waterhammer equations in a latticework of piping elements (Streeter and Wylie, 1968). Examples included two-dimensional pressure waves impinging on arbitrary cylinders, for example see Figure 8, and pressure transients resulting from an explosion in a tank containing liquid with a free surface and a suspended solid body.

Soil Dynamics. Streeter and his colleagues applied the MOC to wave propagation related to earthquake wave motion in unsaturated and saturated soils. One- and two-dimensional wave transmissions were considered (Streeter, Wylie and Richart, 1974), taking into account viscoelastic and strain-softening properties of the soil materials. A two-dimensional latticework simulation of an earth dam subjected to earthquake forces was simulated, leading to potential zones of liquefaction, Figure 9. C. Papadakis (1973, Appendix) and C-P Liou (1976, Appendix) wrote dissertations on this topic.

Pulsatile Blood Flow. The versatility of the waterhammer equations and MOC was demonstrated by analyzing blood flow in arteries (Streeter, Keitzer, and Bohr, 1963, 1964). The relations for flexible rubber-like wall material were combined with the continuity and momentum equations to predict pulsatile patterns of pressure and flow in a canine artery, Figure 10. Two doctoral students who contributed numerically and experimentally to these efforts were W. Zielke (1966, Appendix) and D. Wiggert (1967, Appendix).

Outcome

The accounting of Professor Streeter's professional accomplishments listed herein is not exhaustive, but serve to provide a window into his career as an engineer and educator. Indeed his technical interests have gone far beyond waterhammer. Figure 11 shows his publications in two-year increments, with an indication of topics, to demonstrate the breadth of his contributions.

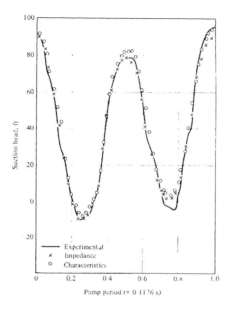

Figure 7. Suction-flange pressure fluctuations for a triplex pump (Wylie and Streeter, 1966).

Figure 8. Pressure pulse around an arbitrary two-dimensional body (Streeter and Wylie, 1968).

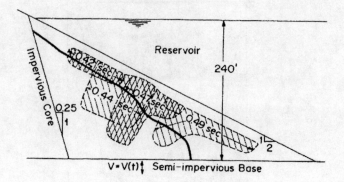

Figure 9. Potential liquefaction in an earth dam (Streeter, Wylie, and Richart, 1974).

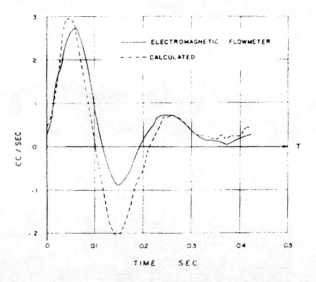

Figure 10. Flow in a canine femoral artery (Streeter, Keitzer, and Bohr, 1963).

Professor Streeter has recalled the advice provided by Hunter Rouse at the end of a short course that he was attending at the University of Iowa in the summer of 1941: "You should get into the study of waves." Water hammer is certainly a subset of the topic, and it was most fitting that he was invited to present the Third ASCE Hunter Rouse Engineering Lecture in 1982, following Donald Harleman, MIT, and John F. Kennedy, Iowa.

In 1974, he was awarded the Worcester Reed Warner Medal from ASME. If one notes some of the outstanding names of earlier winners of this award it is easy to recognize the level of esteem held by Professor Streeter in the eyes of his immediate peers. These names include Timoshenko, Den Hartog, Keenan, Shapiro, and Crandall, all international authorities in their respective fields

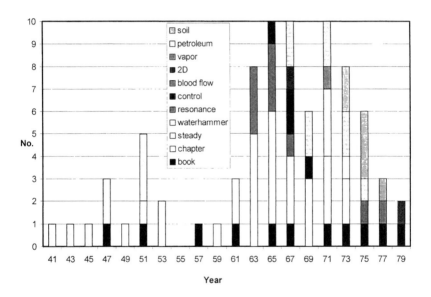

References

Fashbaugh, R.H., and Streeter, V.L. (1965). "Resonance in liquid rocket engine systems," *Journal of Basic Engineering*, ASME, 87 (4), 1011.

Streeter, V.L. (1942). "The kinetic energy and momentum correction factors for pipes and open channels of great width," *Civil Engineering*, 12 (4).

Streeter, V.L. (1950). "Steady Flow in Pipes and Conduits," Chapter 6 in *Engineering Hydraulics*, ed. H. Rouse, John Wiley & Sons, New York, 387-443.

Streeter, V.L. (1963). "Valve stroking to control water hammer," *Journal of the Hydraulics Division*, ASCE, 89 (HY2), 39-66.

Streeter, V.L., Keitzer, W.F., and Bohr, D.F. (1963) "Pulsatile pressure and flow through distensible vessels," *Circulation Research*, XIII (1), 3-20.

Streeter, V.L., and Lai, C.L. (1963). "Water-hammer analysis including fluid friction," *Transactions*, ASCE, 128 (1), 1491-1552.

Streeter, V.L. (1964). "Waterhammer analysis of pipelines," *Journal of the Hydraulics Division*, ASCE, 90 (HY4), 151-172.

Streeter, V.L., Keitzer, W.F., and Bohr, D.F. (1964). "Energy dissipation in pulsatile flow through distensible, tapered vessels," *Pulsatile Blood Flow*, ed. E.O. Attinger, McGraw-Hill, New York, 149-177.

Streeter, V.L. (1965-66). "Computer solution of surge problems," *Proceedings Institution of Mechanical Engineers*, 180 (3E), 62-82.

Streeter, V.L., and Wylie, E.B. (1966). "Hydraulic transients caused by reciprocating pumps," Paper 66-W A/FE-20, ASME, Washington, DC.

Streeter, V.L. (1967a). "Valve stroking for complex piping systems," *Journal of the Hydraulics Division*, ASCE, 93 (HY3), 81-98.

Streeter, V.L. (1967b). "Water-hammer analysis of distribution systems," *Journal of the Hydraulics Division*, ASCE, 93 (HY5), 185-201.

Streeter, V.L., and Wylie, E.B. (1968). "Two and three-dimensional fluid transients," *Journal of Basic Engineering*, ASME, 90 (4), 501-510.

Streeter, V.L. (1969). "Water hammer analysis," *Journal of the Hydraulics Division*, ASCE, 95 (HY6), 1959-1972.

Streeter, V.L., and Wylie, E.B. (1970). "Natural gas pipeline transients," Paper SPE 2555, 44th Annual SPE Fall Meeting, Denver, CO, Sept. 28 – Oct. 1, 1969.

Streeter, V.L., Wylie, E.B., and Richart, F.E. Jr. (1974). "Soil motion computations by characteristics method," *Journal of the Geotechnical Engineering Division*, ASCE, 100 (GT3), 247-263.

Streeter, V.L., Wylie, E.B. (1975). "Transient analysis of offshore loading systems," *Journal of Engineering for Industry*, ASME, 97 (1), 259-265.

Streeter, V.L., (1982). "Transient Cavitating Pipe Flow," Third Hunter Rouse Hydraulic Engineering Lecture, ASCE Hydraulics Division Specialty Conference, Jackson, MS, August 17th.

Wylie, E.B., and Streeter, V.L (1965). "Resonance in Bersimis No. 2 piping system," *Journal of Basic Engineering*, ASME, 87 (4), 925-931.

Wylie, E.B., Streeter, V.L., and Stoner, M.A. (1974). "Unsteady natural gas calculations in complex piping systems," *Journal Society of Petroleum Engineers*, February, 35-43.

APPENDIX—Doctoral students of Professor Streeter at the University of Michigan

Spengas, Aristokolis C. (1959). "An Investigation of Corner Eddies and Free-Surface Instability", co-advisor C.S. Yih.
Lai, Chintu (1962). "A study of waterhammer including effect of hydraulic losses."
Wiggert, James M. (1962). "The development of disturbances in supercritical flows", co-advisor E.F. Brater.
Mahdaviani, Mohammad Ali (1963). "A study of free-surface unsteady gravity flow through porous media."
Contractor, Dinshaw N. (1963). "The effect of minor losses on waterhammer pressure values."
Wylie, E. Benjamin (1964). "Resonance in pressurized piping systems."
Zielke, Werner (1966). "Frequency dependent friction in transient pipe flow."
Baltzer, Robert A. (1967). "A study of column separation accompanying transient flow of liquids in pipes," co-advisor E.F. Brater.
Wiggert, David C. (1967). "Unsteady flows in distributed-outflow Systems."
Stoner, Michael A. (1968). "Analysis and control of unsteady flows in natural gas piping systems," co-advisor E.B. Wylie.
Weyler, Michael (1969). "An investigation of the effects of cavitation bubbles on the momentum loss in transient pipe flow," co-advisor P. Larson.
Propson, Thomas P. (1970). "Valve stroking to control transient flows in liquid piping systems."
Yow, Wu-Shong (1971). "Analysis and control of transient flow in natural gas piping systems," co-advisor E.B. Wylie.
Papadakis, Constantine N. (1973). "Soil transients by characteristics method," co-advisor E.B. Wylie.
Liou, Chyr Pyng (Jim) (1976). "A numerical model for liquefaction in sand deposits," co-advisor E.B. Wylie.

A Tribute to Carl Kindsvater and the Georgia Tech Hydraulics Laboratory

Terry W. Sturm[1], C. Samuel Martin[2]

Abstract

This paper reviews the career of Carl E. Kindsvater who founded the Hydraulics Laboratory at Georgia Institute of Technology. Kindsvater was the recipient of two ASCE Norman medals, the ASCE Collingwood Prize, the Rickey Medal, the Julian Hinds Award, and was a recognized authority on open-channel flow. He is especially known internationally for seminal work on weirs, flow over highway embankments, flow through bridge openings, and hydraulic jumps in sloping channels. Photos of equipment and various experiments reveal his procedures and methods.

Figure 1. Carl Edward Kindsvater (1913-2002)

Introduction

Carl Edward Kindsvater was born in Kansas in 1913, a second-generation child of German ancestry. His grandfather Johann Heinrich Kindsvater (1844-1893) immigrated

---

[1]Professor, Georgia Institute of Technology. e-mail: terry.sturm@ce.gatech.edu

[2]Professor Emeritus, Georgia Institute of Technology. e-mail: smartin@ce.gatech.edu

to Russell, Kansas in 1892, settling in a community of distinctive Germans. The people of this community emigrated from the Lower Volga in Russia, causing them to be referred to as Volga Deutsch; yet they had lived outside of Germany for nearly two centuries. In fact, their forefathers were invited to Russia by Catherine the Great in 1766 to settle in the rich farmlands of the Lower Volga. These Germans were also known as the Czar's Germans. Actually, after his retirement Kindsvater did extensive research into the German-Russian connection.

Educational Background

Carl Kindsvater matriculated at the University of Kansas, graduating with a bachelor's degree in civil engineering in 1935. He then enrolled at the University of Iowa to pursue graduate study in hydraulics. Following him by one year at both Kansas and Iowa was his friend John S. McNown (1916-1998), who later became a renowned professor at the latter university. Kindsvater received his M. S. degree in 1936. His thesis on the characteristics of a hydraulic jump in an enclosed circular conduit was supervised by Professor E. W. Lane. Part of this work was later published as a note by Lane and Kindsvater (1938).

Subsequent to graduation from Iowa in 1937, Kindsvater accepted employment at the Norris Laboratory at the Tennessee Valley Authority (TVA), where he was involved in various model studies and other experimental investigations. Following the death of David L. Yarnell, a U.S. Department of Agriculture (USDA) employee stationed at Iowa, Kindsvater extended his investigation of the characteristics of hydraulic jumps to the effect of sloping channels using data collected by Yarnell at Iowa. Kindsvater then, as part of his responsibilities, pursued an analysis that culminated in a meaningful paper (Kindsvater, 1944), which was awarded the Collingwood Prize in 1945.

Hydraulic Laboratory Design and Early Career at Georgia Tech

In order to gain practical experience Kindsvater left TVA to work for the Corps of Engineers at the Little Rock District. In 1945 he received a call to establish a hydraulics laboratory and graduate program at the Georgia Institute of Technology. Although Georgia Tech had no hydraulics laboratory, it is noteworthy that the president of the institute was Colonel Blake Van Leer (1893-1956). Van Leer was a graduate of Purdue University and the University of California, Berkeley, and more significantly, an ASME Freeman Scholar in 1927-28, and the developer of the so-called California-Pipe method of flow metering, based upon the brink depth at culvert exits. Van Leer was instrumental in attracting Kindsvater to Atlanta, lending support for establishing a viable hydraulics laboratory and academic program at an institution that was primarily an undergraduate technical school at the time, albeit with a strong regional and even national reputation. Indeed, the authority to award the doctorate was only approved in 1946. During Kindsvater's earliest days at Georgia Tech, he had complete enthusiastic support from Dean of Engineering Mason, President Van Leer and numerous powerful alumni. However, these were not easy times. Kindsvater himself stated in 1988: *"For six years it was a one man operation, starting from absolute zero (When I arrived: no laboratory facilities; a weak CE department with no faculty interest in hydraulics). But in that six-*

year period, hydraulics and hydraulics research at Georgia Tech attracted widespread favorable notice--not to mention candidates for additions to the staff." (Personal communication)

Major Equipment. When Kindsvater arrived at Georgia Tech the only existing elements of a laboratory were two crates containing pumps, which had been acquired some five years earlier to supply a constant head tank. Kindsvater, with his intense energy, drive, forbearance, and ability, embarked on a program of planning, design, and construction of a well-equipped hydraulics laboratory. For the first five years the laboratory was rather small, but adequate for positioning various facilities, all linked to a well-designed constant head tank. With limited resources he managed via donations and other means to render the laboratory operational within two years. This was in no small part accomplished with the able assistance of a master craftsman named Homer Bates, who by some divine decree, simply walked into the lab in 1947. The pairing of these two minds bore abundant fruitful results for Georgia Tech. Within two years (1947) the major components that had been constructed were (1) a re-circulating water system with a constant-head tank (Figure 2) that could supply some 20 feet of head, (2) an automatic weighing tank (Figure 3) with capacity of 10,000 lbs, and (3) a 25-ft long 3-ft wide glass-walled flume (Figure 4) for research and model studies of spillways and other structures. This flume has been a work horse for the laboratory for over 50 years, being used for some 20-30 investigations. Initially, Kindsvater utilized the flume for weir studies and flow over highway embankments. For hydraulic model investigations and fundamental research into flow through bridge constrictions, for which the USGS was a stalwart partner, he supervised the construction of a 10-ft wide horizontal flume (Figure 5). Much work was accomplished with these two flumes before the laboratory was expanded in 1953.

Kindsvater was given the authority to occupy the entire basement of the Civil Engineering Building, which had a floor space of approximately 100 ft by 40 ft. By this time he had a feasible and effective working relationship with the USGS as a consultant, advisor, mentor, and professor. The USGS and Georgia Tech designed and constructed two lasting facilities, a 42-inch wide 80-ft long tilting flume, and a 14-ft wide 60-ft long horizontal basin. These two facilities, which are still in active usage, formed the foundation of intensive and lasting research in open-channel hydraulics. It was clear that he subscribed to Hunter Rouse's classic concept of "engineering hydraulics" as including fluid mechanics.

Instrumentation. It should be mentioned that Kindsvater, operating with a limited budget, improvised remarkably by rendering pipe elbows (Figures 6 and 7) and pipe reducers (Figure 8) into flow meters, albeit realizing that flow calibration was a necessity. Hence, the importance and reliance on the constant-head tank to meter flow directly entering or leaving the flumes, or for direct calibration of bend meters and so-called Venturis, which were comprised of (a) a sudden reduction to the meter throat, followed by (b) a gradual expansion by another pipe reducer turned around. Kindsvater and master technician Bates designed and fabricated not only Pitot tubes, but also Pitot-static tubes on the geometry proposed by Prandtl, both of which can be seen in Figure

9. Aspects of simplifying Pitot-tube design are suggested in Kindsvater (1947).

Instruction. The 15-inch wide glass-walled flume with sluice gate and tail gate shown in Figure 10, was designed by Kindsvater in 1948 to be used mainly for instructional purposes. This flume, for which one can measure pressure distribution on the sluice gate and on the floor, is still in use today for undergraduate and graduate level demonstrations and experiments. At a period in time when electronic instrumentation was not used commonly, care was taken to construct accurate manometers, hook gauges, and precision micro manometers for pressure measurement. An example of a multiple-tube manometer board is illustrated in Figure 11. Both undergraduate and graduate instruction was important to Kindsvater, who devised various simple and often ingenious experiments. Although his main field was free-surface flow, he supervised M.S. work on aspects of flow in enclosed conduits; for example, Fleetwood (1955) and Kittle (1956). Figure 12 shows a portion of an instrumented pipe-friction experiment. A worthwhile experiment for instruction, still in use today, is the measurement of piezometric and total head distribution through both an orifice and a Venturi meter by means of twelve pressure taps, shown on Figure 13. Figure 14 shows Kindsvater (far right) instructing students on using a current meter at a weir in the 3-ft flume. His interest in cavitation may well have stemmed from his observation of sonoluminescence at the exit of draft tubes, Kindsvater (1945). Figure 15 shows a cavitation rig built for exploratory purposes.

Hydraulic Model Studies. Beginning in 1949, Kindsvater had already convinced various power companies in adjoining states that hydraulic model studies were essential for some of their existing and new projects. Over the next 6-8 years several comprehensive model studies of spillway and tailrace structures were conducted, two of which are shown in Figures 16 and 17. He also utilized the 3-ft glass-walled flume for sectional models, as shown for example in Figure 18. He received the ASCE Rickey Medal in 1955, in part due to his contribution to model studies reported in Harrison and Kindsvater (1954). He contributed to the understanding of draft-tube performance for Martin Dam Power Plant, Alabama Power. An overall view of the scaled model is illustrated in Figure 19, along with him performing measurements in Figure 20, reported in Kindsvater and Randolph (1955).

Classic Papers

The enduring legacy of Carl Kindsvater's contributions to hydraulic engineering can be gauged not only by the professional awards that he received for papers summarizing his research, but also by the continuing influence of the results of his research in commonly accepted modern hydraulic analyses and computer models. His papers have the hallmark of a careful experimentalist with deep physical insights into the fluid mechanics of the phenomenon being investigated, but they are always completed with an eye toward the practical engineering applications of his work. Kindsvater's papers on experimental investigations each begin with an insightful dimensional analysis and a thorough explanation of the physics involved to isolate the most important dimensionless parameters. The experiments are exhaustive (and probably were exhausting to his students) with a final result in dimensionless form that is immediately applicable to engineering practice. His achievements in experimental hydraulics are all the more remarkable considering his arrival in 1945 at Georgia Tech, where no hydraulics lab existed at the time, and the completion of a classic body of hydraulics research in the ensuing 15 years. Four of his classic contributions are summarized here in historical order, and their lasting significance is documented.

The first classic paper published by Carl Kindsvater to be highlighted here is the one entitled "The hydraulic jump in sloping channels" (Kindsvater 1944). It is included not only because it won ASCE's Collingwood Prize but also because it bridges the crucial transition in Kindsvater's career beginning from new graduate of the University of Iowa with an M.S. degree in 1937, through his practice-oriented work with the Tennessee Valley Authority (TVA) and the U.S. Army Corps of Engineers (COE), and leading up to his arrival at Georgia Tech in 1945. Kindsvater's M.S. thesis, supervised by E. W. Lane at Iowa, was entitled "Hydraulic jump in enclosed conduits" and was duly published (Lane and Kindsvater 1938). However, of more interest here is that Kindsvater assisted D. L. Yarnell at Iowa in the collection of data on hydraulic jumps in sloping open channels from August 1936, after he finished his M.S. thesis, until January 1937 when he presumably began his career with TVA. Yarnell was a senior engineer for the U.S. Department of Agriculture (USDA) and was working at Iowa at the time. (Yarnell is also well known for his experimental work on backwater caused by bridge piers). Shortly after Kindsvater left Iowa, Yarnell died in March 1937, and nothing further was done with the hydraulic jump data. Subsequently, Kindsvater embarked on an investigation of spillway design at TVA and revived the data on hydraulic jumps in sloping channels collected at Iowa apparently with the blessing of the USDA.

In the paper, he first classified the hydraulic jump on sloping channels in terms of the position of the jump relative to the break in slope between the sloping channel and a horizontal stilling basin. He then applied the classic momentum analysis but with the inclusion of what amounts to a weight term resolved down the slope which involves an empirical factor determined by experiment. The empirical factor was added in an ingenious way so that the solution for the sequent depth is of the same mathematical form as for the hydraulic jump in a horizontal channel. In a subsequent discussion of the paper by G. H. Hickox of TVA, more extensive data collected at TVA were shown to agree with the analysis offered by Kindsvater for other slopes. The most telling aspect of the paper was Kindsvater's thorough closure response to the extensive discussion of the paper. In answering a particular criticism that the hydraulic jump on a slope was really a submerged jump rather than being similar to the jump in a horizontal channel, Kindsvater devised a small channel in his bathtub at home and produced some beautiful photographs that clearly supported his view. The closure to the paper was published in 1944, just one year before he was attracted to Georgia Tech from the COE office in Little Rock, Arkansas. Kindsvater's closure to the paper clearly demonstrated that he had the ability for a successful academic career. This particular paper by Kindsvater has become a classic part of the hydraulic engineering literature. It is cited by Chow (1959) and featured in the comprehensive treatments by Bradley and Peterka (1957) and by Rajaratnam (1967) which are primary references in most subsequent relevant publications.

The second classic paper by Kindsvater to be discussed here is the one entitled "Tranquil flow through open channel constrictions" (Kindsvater and Carter 1955) which won the 1956 ASCE Norman Medal. It was also chosen as a classic paper by an ASCE Task Committee chaired by John McNown on classic papers in hydraulics from 1935-1960. (ASCE 1982). The motivation for this research came from the U. S. Geological Survey (USGS) to use bridge contractions for indirect measurement of discharge, but it has received far wider application in bridge hydraulics. Kindsvater and Carter conceptualized the flow through a bridge contraction as being similar to a pipe orifice problem but in a free surface flow. Accordingly, they wrote the energy and continuity equations from an approach section of the bridge to the contracted section just downstream of the bridge opening (i.e., the *vena contracta*) including entrance losses and friction losses. They then developed a dimensional analysis for the resulting coefficient of discharge and methodically produced a set of experimental curves for predicting the coefficient as a function of the geometric contraction ratio, the ratio of bridge flow length to opening length, Froude number, and degree of entrance rounding. Kindsvater and Carter recognized that for widespread acceptance and application of the methodology, it had to take into account characteristics of the flow field dictated by compound channel flow in natural channels. As a result, they introduced a discharge contraction ratio, m, which has received widespread use in bridge hydraulics modeling. It was defined as the ratio of the discharge blocked by the bridge abutments and embankments in the bridge approach flow to the total discharge, and $(1 - m)$ replaced the simple geometric contraction ratio. Using the discharge contraction ratio, Kindsvater and Carter developed the coefficient of discharge curves in the laboratory for several compound channel geometries with varying roughnesses. Results for the discharge coefficient are shown in Figure 21, and the experimental setup can be seen in Figures 22 and 23. Portions of this work first appeared in a Geological Survey Circular (Kindsvater and Carter 1953) which was later updated by Matthai (1967) and then became a fundamental building block in current computer models of bridge hydraulics, namely WSPRO (Shearman *et al.* 1986) and HEC-RAS (Corps of Engineers 1998).

The third classic paper by Kindsvater and Carter (1959) is perhaps as close as possible to the last word on that staple of hydraulic engineering research for over two centuries, the sharp-crested weir. This paper received the ASCE Norman Medal in 1960. Kindsvater and Carter's insightful dimensional analysis isolated the viscous and surface tension effects from the purely geometric influences on the weir discharge coefficient for rectangular notch weirs as well as suppressed weirs used for the measurement of water flow. They expressed these influences as additive corrections to the head and crest length of the weir. Thus the discharge coefficient depended only on the geometric ratio of the head on the weir crest to the height of the weir, and on the geometric ratio of the weir crest length to the approach channel width. This is illustrated in Figure 24 in which the corrections to the head and crest length are shown to produce a relationship for the weir discharge coefficient that is valid for both small and large values of the head and crest length. Sharp-crested weir flows in the laboratory are shown in Figures 14 and 25. Kindsvater and Carter found reasonable agreement with a Swiss weir formula used for verification but concluded in general that the scatter in previous investigators' results is largely due to differences in weir geometry and particularly to differences in the approach velocity distribution. Although Kindsvater and Carter despaired of any universal, precise weir formula, their formula has come to be accepted as a standard in many hydraulics handbooks such as Brater *et al.* (1996) in its 7th edition and Bos (1988). Quoting Brater *et al.*, the procedure of Kindsvater and Carter "provides what is probably the best available method for estimating the discharge through rectangular notched sharp-edge weirs."

The final classic paper to be considered here, "Discharge characteristics of embankment-shaped weirs", is actually a USGS water supply paper (Kindsvater 1964). It represents the collection of data by graduate students and USGS assistants in the Georgia Tech hydraulics laboratory over nearly a decade on a problem of highway embankment overflow that remains important today in bridge and culvert hydraulics. Although it did not receive an award, it is emblematic of Kindsvater's hydraulics research program at Georgia Tech in its comprehensiveness, its blend of theory and experiment, its engineering applicability, and its extensive involvement of graduate students and USGS employees. It also represents Kindsvater's last major work in hydraulic engineering before he made a career change to water resources engineering in the early 1960s. The paper summarizes the results of experiments on embankment models in the 3-ft wide, glass-walled flume in the Georgia Tech hydraulics laboratory as shown in Figure 26. Kindsvater treats the embankment like a broad-crested weir as illustrated in Figure 27 and includes the effects of varying embankment geometry, embankment roughness, and tailwater submergence to derive relationships for the discharge coefficient. In the upper photograph of Figure 28 is shown the case of free plunging flow at the upper limit of the transition to submerged flow over the embankment, and in the lower photograph is shown the switch to free, surface flow just prior to submergence by the tailwater. The relationships for discharge coefficient have subsequently become the basis for calculating embankment overflow for bridges and culverts in widely accepted computer models in the U.S. They can be found in the Federal Highway Administration publications of HDS-1, "Hydraulics of Bridge Waterways" (Bradley 1978); HDS-5, "Hydraulic Design of Highway Culverts" (FHWA 1985); and in the methodology for the FHWA/USGS computer program for bridge hydraulics, WSPRO (Shearman *et al.* 1986).

Establishment of Water Resources Center

Kindsvater devoted the latter portion of his career to interdisciplinary water resources research and education. After a sabbatical to Stanford and Berkeley in 1961-62, he returned to Georgia Tech to establish a water resources center in 1963. It was approved by the Board of Regents of the University System of Georgia "to facilitate and coordinate a broad-based, campus-wide program of water resources education and research at the Georgia Institute of Technology" (Water Resources Center, 1966). Kindsvater had a clear vision of an independent research center with its own staff, budget, and interdisciplinary degree-granting status. As a result, he was uniquely positioned for the establishment of the Center to administer the federal water resources research program through the Office of Water Resources Research (OWRR) under the provisions of the Water Resources Research Act of 1964 (Public Law 88-379). This required an act of the Georgia General Assembly passed in 1965 which is all the more remarkable because Georgia Tech is not the land-grant university in Georgia. However, Kindsvater formed a joint advisory board for the Center with equal representation from the University of Georgia and involved many University of Georgia researchers in the Center's research activities.

In 1966, the Center was detached from the budget and administration of the School of Civil Engineering and the College of Engineering and attached to the Office of the Vice President for Academic Affairs in keeping with Kindsvater's desire to make the Center an independent entity. Kindsvater extended the reach of the Center far beyond the OWRR funding. It also administered student training grants from the Federal Water Pollution Control Administration (FWPCA) and later the Environmental Protection Agency (EPA). He established cooperative relationships with the Army Corps of Engineers and the Agricultural Research Service. In addition, he maintained a continuing hydraulic research contract with the USGS as a part of the Center.

In 1970, the name of the Center was changed to the Environmental Resources Center and its activities involved nearly 30 research projects with a professional staff of seven salaried members including professors in water resources planning, economics, and social aspects of water resources projects. There were 11 graduate students who were enrolled through cooperative programs as employees of the COE and the USGS, and another 9 graduate students were supported on federal training grants. The Center continued to coordinate graduate programs and courses in water resources planning from departments all across the Georgia Tech campus.

By 1972, Kindsvater had decided to move on to become a senior scientist for the USGS, a position from which he retired in 1976, but he had clearly spent the last 10 years of his career at Georgia Tech successfully realizing his vision of interdisciplinary water resources research and education administered by the water resources center. He overcame institutional and political obstacles to lead the education of a new generation of water resources scientists and engineers. His full views on water resources education are detailed elsewhere (Kindsvater 1964) and are still quite applicable. In his last year as director of the Center, he wrote (Environmental Resources Center, 1972):

"*The Center was conceived in optimism regarding Georgia Tech's ability to respond to the need for problem-focused, Institute-wide programs related to natural resources management. The low-profile coordinative role assumed by the Center has proved to be a vulnerable, often a frustrating one. Nonetheless, on the assumption that men of good will can rise above the rigidities, inertia, and internal rivalries which characterize so many universities, the Georgia Tech Environmental Resources Center is believed to represent an effective, least-cost method of fostering environmental resources education and research.*"

Indeed, the prominence of interdisciplinary research centers all across today's Georgia Tech campus as well as at many other universities are testament to Kindsvater's vision of cutting across traditional university disciplines to attack and solve complex societal problems. In recognition of his many years of service to the field of water resources development and his general contributions, ASCE awarded Kindsvater the Julian Hinds Award in 1979.

ASCE Activities

Carl Kindsvater was remarkably active in ASCE activities, both technically and professionally. He served on various Hydraulics Division technical committees, being elevated to the Executive Committee, as well as hosting the annual Hydraulics Conference at Georgia Tech in 1958. On the professional side he was president of the Georgia Section of ASCE, and he served as director of District 10 on ASCE's Board of Direction from 1963 - 1966. In 1981 he was honored by being appointed as an honorary member.

Legacy

How does one summarize the legacy of such a fruitful career as that of Carl Kindsvater, or really two careers, one in hydraulic engineering and another in water resources research? Certainly, his energy, vision, intellectual ability, and the sheer force of his personality brought much respect, not only to himself but also to Georgia Tech. He is also highly regarded by the USGS for his many contributions to their mission. He won numerous professional awards for his research and service. He left behind an excellent operational hydraulics research laboratory, undergraduate instructional laboratory, and a water resources research center, all of which have continued to thrive at Georgia Tech. His classic papers form the bedrock of perhaps the most formative and influential era of hydraulic engineering. He educated a new generation of scientists and engineers in hydraulics and water resources who remember him fondly as an outstanding teacher, "fair but tough as nails" (personal communication, Harry Barnes). By all measures, his career is worthy to be honored in the classical tradition of Henry Darcy.

References

ASCE. (1982). *Classic Papers in Hydraulics 1935-60*, prepared by the Task Committee to Develop a Volume of Classics in Hydraulics of the Hyd. Div., ASCE, 488-525.

Bos, M. G. (1988). *Discharge Measurement Structures*, ILRI Publication 20, 3rd Revised edition, Wageningen, the Netherlands.

Bradley, J. N. (1978). *Hydraulics of Bridge Waterways* (HDS-1), Federal Highway Administration, Washington, D.C.

Bradley, J. N., and Peterka, A. J. (1957). "Hydraulic Design of Stilling Basins: Stilling Basin with Sloping Apron (Basin V)," *J. Hydr. Div.*, ASCE, vol. 83, no. HY5, Paper 1405, 1-32.

Brater, E. F., King, H. W., Lindell, J. E., and Wei, C. Y. (1996). *Handbook of Hydraulics*, 7th edition, McGraw-Hill, New York.

Chow, V. T. (1959). *Open Channel Hydraulics*, McGraw-Hill, New York.

Corps of Engineers (1998). *HEC-RAS Hydraulic Reference Manual*, Hydrologic Engineering Center, U.S. Army Corps of Engineers.

Environmental Resources Center (1972). *Annual Report*, Georgia Institute of Technology, Atlanta.

Federal Highway Administration (FHWA) (1985). *Hydraulic Design of Culverts*, Report FHWA-IP-85-15, (HDS-5), Federal Highway Administration, U. S. Dept. of Transportation, Washington, D.C..

Fleetwood, Trafton Webb (1955). "Abrupt Enlargements in Smooth and Rough Pipes", MS Thesis, Georgia Tech, 62 pages.

Harrison, E. S. and Kindsvater, Carl E. (1954). "Dam Modifications Checked by Hydraulic Models", *TRANSACTIONS ASCE,* Vol. 119, 73-92.

Kindsvater, Carl E. (1936). "The Hydraulic Jump in Enclosed Conduits", *M. S. Thesis,* State University of Iowa.

Kindsvater, Carl E. (1944). "The Hydraulic Jump in Sloping Channels", *TRANSACTIONS ASCE,* Vol. 109, 1107-1120 (Discussions by C. J. Posey and J. C. Stevens).

Kindsvater, Carl E. (1945). "Phenomenon of 'Flashes' at Outlet of Sluices", *Civil Engineering, ASCE,* Vol. 15, 565.

Kindsvater, Carl E. (1947). "Simplified Designs Facilitate Pitot Tube Application to Small Pipes", *Civil Engineering, ASCE,* November, 680.

Kindsvater, Carl E. and Carter, R. W. (1953). "Computation of Peak Discharge at Contractions", *U. S. Geological Survey Circular No. 284.*

Kindsvater, Carl E. and Carter, R. W. (1955). "Tranquil Flow through Open-Channel Constrictions", *TRANSACTIONS ASCE,* Vol. 120, 955-980 (Discussions by E. M. Laursen and A. Toch, F. W. Blaisdell, P-N. Lin, C. F. Izzard, and C. J. Posey).

Kindsvater, Carl E. and Randolph, R. R. (1955). "Hydraulic Model Studies of Martin Draft Tubes", *TRANSACTIONS ASCE,* Vol. 120, 1399-1419 (Discussion by R. S. Woodruff).

Kindsvater, Carl E. and Carter, R. W. (1959). "Discharge Characteristics of Rectangular Thin-Plate Weirs", *TRANSACTIONS ASCE,* Vol. 124, 722-822 (Discussions by T. Sarpkaya, S. Kolupaila, R. W. Powell, J. W. Paull, I. Oki, and M. R. Carstens).

Kindsvater, Carl E. (1964a). "Discharge Characteristics of Embankment-Shaped Weirs", *U. S. Geological Survey Water-Supply Paper 1617-A,* 114 pages.

Kindsvater, Carl E. (1964b). "Trends in Water Resources Education", *Journal of Professional Practice, ASCE,* PP2, May, 3889.

Kittle, Benjamin Lee (1956). "Abrupt Enlargements in Circular Pipes", MS Thesis, Georgia Tech, 59 pages.

Lane, E. W., and Kindsvater, Carl E. (1938). "Hydraulic Jump in Enclosed Conduits", *Engineering News-Record,* December 29.

Matthai, H. F. (1967). "Measurement of Peak Discharge at Width Contractions by

Indirect Methods," *Techniques of Water Resources Investigations*, Chapter A4, Book 3, USGS.

Rajaratnam, N. (1967). "Hydraulic Jumps," *Advances in Hydroscience*, vol. 4, Academic Press, New York, 197-280.

Shearman, J. O., Kirby, W. H., Schneider, V. R., and Flippo, H. N. (1986). *Bridge Waterways Analysis Model: Research Report*, Report FHWA/RD-86/108. Federal Highway Administration, Washington D.C.

Water Resources Center (1966). *Annual Report*, Georgia Institute of Technology, Atlanta.

Appendix – Additional References

Bravo Restrepo, Pablo (1960). "Influence of Fluid Properties Related to Thickness of Weir Crest", MS Thesis, Georgia Tech, 68 pages.

Carter, Rolland William (1956). "A Comprehensive Discharge Equation for Rectangular-Notch Weirs", MS Thesis, Georgia Tech, 72 pages.

Davidian, Jacob (1959). "Influence of the Boundary Layer on Embankment-Shaped Weirs", MS Thesis, Georgia Tech, 97 pages.

Ferguson, G. E. *et al.* (1957). "A History of the Water Resources Division", U. S. Geological Survey, Vol. V, July 1, 1947 June 30, 1957.

Hudson, H. H., Cragwall, J. S., *et al.* (1966). "A History of the Water Resources Division", U. S. Geological Survey, Vol. VI, May 1, 1957 - June 30,1966.

Jones, Jack Carter. (1954). "Influence of Several Geometric Variables on the Efficiency of Two Modern Turbine Draft Tubes", MS Thesis, Georgia Tech, 54 pages.

Kindsvater, Carl E. (1961). "Energy equation for Partially Developed Free-Surface Flow", *Civil Engineering, ASCE,* March, 66.

Kindsvater, Carl E. (1964-65). "Organization and Methodology for River Basin Planning", Proceedings, Water Resources Center, Georgia Institute of Technology, Two Volumes.

Prawel, Sherwood Peter (1958). "Discharge Characteristics of an Embankment-Shaped Weir", MS Thesis, Georgia Tech, 58 pages.

Schnabel, George Benjamin (1953). "Energy Loss Due to Diametric Cylindrical Obstructions in Pipes", MS Thesis, Georgia Tech, 30 pages.

Sigurdsson, Gunnar (1955). "Discharge Characteristics of an Embankment-Shaped Weir", MS Thesis, Georgia Tech, 83 pages.

Wells, James Robert (1953). "Discharge Characteristics of Rectangular Notch Weirs in Rectangular Channels", MS Thesis, Georgia Tech, 39 pages.

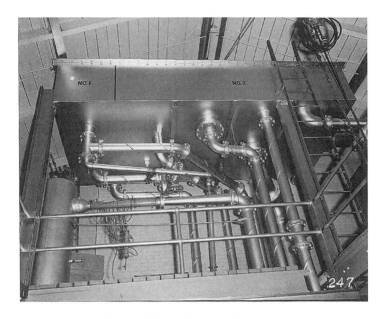

Figure 2. View of Constant-Head Tank

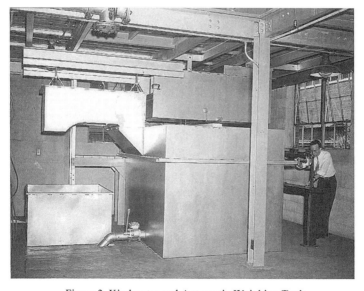

Figure 3. Kindsvater and Automatic Weighing Tank

Figure 4. View of 3-ft Wide Glass-Walled Flume

Figure 5. View of 10-ft Flume

Figure 6. View of 6"-6" Bend Meter

Figure 7. View of 6"-3" Bend Meter

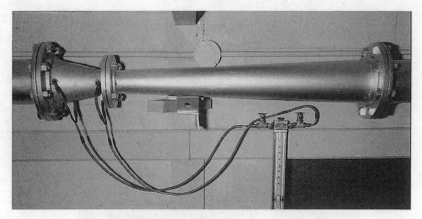

Figure 8. View of Pipe-Reducer Venturi

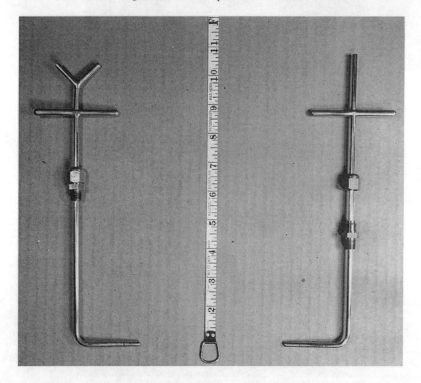

Figure 9. View of Pitot and Pitot-Static Tubes

Figure 10. View of 15-inch Wide Glass-Walled Flume

Figure 11. View of Multiple-Tube Manometer Board

Figure 12. View of Pipe-Friction Apparatus

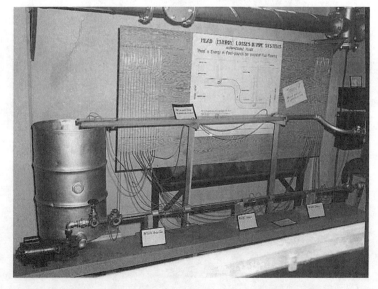

Figure 13. View of Undergraduate Orifice-Venturi Experiment

Figure 14. Kindsvater and Students with Weir in 3-ft Flume

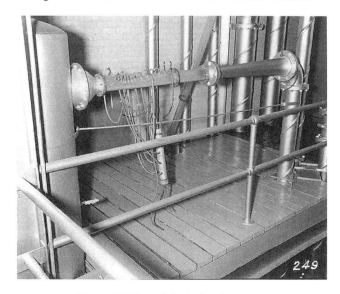

Figure 15. View of Cavitation Apparatus

Figure 16. View of Buttress Spillway Model in 10-Ft Flume

Figure 17. View of Spillway Model in 10-Ft Flume

Figure 18. View of Movable Bed Model of Spillway in 3-Ft Flume

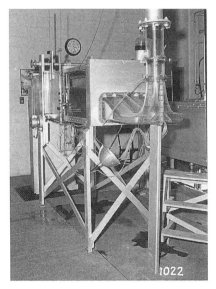

Figure 19. View of Draft Tube Model

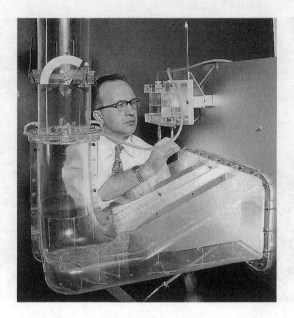

Figure 20. Kindsvater with Model of Martin Dam Draft Tube

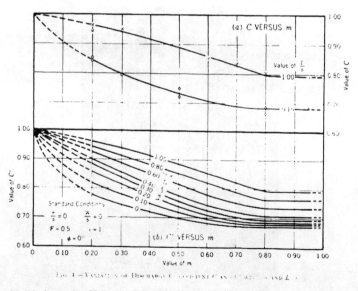

Figure 21. Discharge Coefficient in Open Channel Constrictions (Kindsvater and Carter 955)

Figure 22. View of Bridge Constriction in 10-ft Flume

Figure 23. View of Flow Through Bridge Constriction in 10-ft Flume

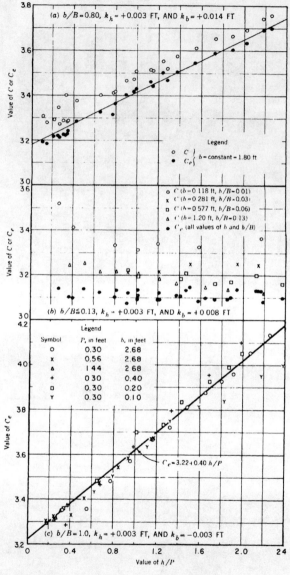

FIG. 6. —GEORGIA TECH TESTS

Figure 24. Sharp-Crested Weir Discharge Coefficient (Kindsvater and Carter 1959)

Figure 25. Flow Over Rectangular Weir in 10-ft Flume

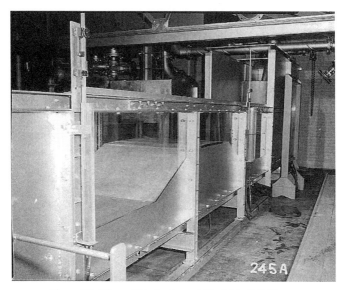

Figure 26. View of Embankment Model in 3-ft Flume

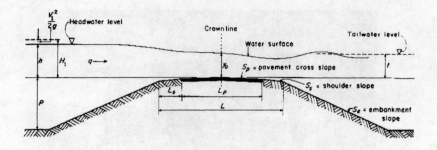

Figure 27. Sketch of Geometry of Highway Embankment Model Study

Figure 28. Free-Surface Shapes for Embankment Flow in 3-ft Flume

Floyd Nagler, Founding Director of IIHR

Cornelia F. Mutel and Robert Ettema[1]

Abstract

Floyd Nagler, the founding director of the Iowa Institute of Hydraulic Research (IIHR, now IIHR-Hydroscience & Engineering), was granted a brief 13 years to leave his imprint on this fledgling institute. Yet in that period, he managed to construct a substantial physical research structure, establish an innovative administrative scheme, and bring considerable stability to IIHR by embracing governmental substations that brought in research projects and financial support. Nagler also imbued IIHR with his spirit of enthusiasm and energy. In these many ways, he laid the groundwork for an institute that would become world-renowned for its contributions to hydraulic education and research, and which continues to reflect the diversity of subject matter and approach modeled by its colorful and multi-talented founder.

Introduction

Many hydraulic engineers associate IIHR with Hunter Rouse, John F. Kennedy, or other present-day notable researchers whose lives have pivoted around this University of Iowa institution. Few realize that IIHR actually was founded by someone similar to Rouse and Kennedy in stature, energy, and determination: Floyd Nagler, whose broad vision and multifaceted talents were equal to the task of laying the groundwork of this now-renowned institution. Who was this man, and what was his legacy to IIHR?

Born in 1892, Nagler grew up in Michigan where his father, a Methodist minister, was fervent in conducting revival meetings and promoting prohibition. His father's calling seemed to imprint Nagler's own life. Many years later, the adult Floyd Nagler would take over the pulpit whenever the Methodist minister of his Iowa City church left town, and Nagler reportedly saw his life's purpose as reconciling science, technology, and religion by explaining the workings of God's creation to the lay public.

Nagler displayed a hunger for hard work from his youth. Starting at age 10, he regularly performed odd jobs for nearby neighbors and businesses. Entering Michigan Agricultural College in 1910, he undertook engineering studies and then proceeded to the University of Michigan, where in 1917 he received his doctorate in

---

[1] Both IIHR-Hydroscience & Engineering, College of Engineering, The University of Iowa, Iowa City, IA, 52242; Mutel – phone 319-335-5315, connie-mutel@uiowa.edu; Ettema – phone 319-335-5224, robert-ettema@uiowa.edu

Hydraulic Engineering. In those scant seven years, in addition to earning his degrees, he also performed field studies for an engineering consulting firm, directed the construction of a University of Michigan hydraulic flume, served there as a Teaching Assistant, and received several fellowships. For three years following his graduation, he worked for private consulting firms in New York and enlisted as a meteorological expert for the U.S. Signal Corps and U.S. Weather Bureau.

Thus by age 28, in 1920, he already had 18 years of diverse and demanding work experience behind him. That year Nagler (Figure 1.) arrived in Iowa City, Iowa, to become Assistant Professor of Mechanics and Hydraulics at the University of Iowa. His task was to manage its newly-constructed Hydraulics Laboratory, built on campus alongside the Iowa River.

Iowa's Hydraulics Laboratory

This new Hydraulics Laboratory, a small structure a mere 22 feet on each side, was one of about four dozen hydraulics laboratories with experimental facilities that were built in the U.S. in the late 1800s and early 1900s. Three-quarters of them were located at educational institutions, with the remainder operated by the government or private companies. These laboratories were a testimony to the concept that hydraulic processes could be better understood by using laboratory models for testing complex,

Figure 1. Floyd Nagler, observing the Mississippi River's flow over Keokuk Dam spillways. (IIHR Archives)

real-life processes. Iowa's Hydraulics Laboratory was constructed with such model use in mind. The laboratory spanned a 10-foot-wide open flume, fed directly by Iowa River water, within which models could be constructed. The flume could be accessed either from outside the building or by lifting the laboratory's floor planks.

During Nagler's tenure at Iowa, the field of hydraulics would rapidly expand in importance and breadth of activity. Hydraulics-related problems were abundant and obvious. As Nagler himself later wrote, "The field open to hydraulic research is almost unlimited." The laboratories popping up across the country were stimulating an attack on basic problems that had not previously been investigated. Spin-off projects were attacked with zest, for hydraulic research was providing useful and necessary answers to basic questions.

Nagler, newly arrived in Iowa, accepted responsibility for a laboratory structure that initially lacked research funding and focus. He first assumed that his research, and that of any students, would be supported in the tradition of the times: by performing contract work for industry. In particular, Nagler intended to establish a turbine-testing facility at Iowa, with private industry providing the necessary monies to keep his laboratory open.

Fate expeditiously led in another direction. Between 1921 and 1923, Nagler was commissioned by the U.S. Department of Agriculture (USDA) to perform tests concerning the flow of water past obstructions. In 1922 David Yarnell, a USDA engineer, arrived at the Hydraulics Laboratory to assist with the studies. He and Nagler quickly became friends as well as colleagues, the two completing many joint research investigations. Yarnell, the first of a sequence of U.S. government employees to be officially stationed at the Hydraulics Laboratory, remained here until a coronary occlusion took his life in 1937. Prior to World War II, governmental grants for university research were largely nonexistent. Yet through the establishment of Yarnell and his USDA substation, the Hydraulics Laboratory essentially became a federally subsidized institution, with funds and applied projects to support students and auxiliary projects. With this support, the Hydraulics Laboratory became one of the first U.S. educational hydraulics laboratories to perform basic as well as applied research.

Yarnell's arrival was instigated by the USDA – Bureau of Roads' need to identify efficient culvert underpasses for the nation's growing highway system. Thus a series of full-scale culverts were constructed in the laboratory's flume, each culvert varying in shape, size, outlet structure, or material (Figure 2.). Sequences of culverts were then tested for the quantity and speed of water's flow, with the results fed directly into highway improvements. Yarnell also collaborated with Nagler on research regarding the scouring of riverbeds around bridge piers of various shapes, the flow of water around river bends, and friction losses in pipe bends.

Thus, while the first research work at the Hydraulics Laboratory was synonymous with Nagler's efforts, he soon took charge of a growing research staff. Nagler however remained involved in the Hydraulics Laboratory's research as well as its administration, and as the years passed he seemed to retain a thorough grasp of IIHR's diverse efforts and their import.

Nagler did not allow his engagement with laboratory experiments to restrict his love of wandering the Iowa countryside and pondering water's impact. In 1928,

Figure 2. Staff engineers studying water's flow through models of culverts of varied sizes, shapes, and materials, a subject that became a major emphasis at the Hydraulics Laboratory in the 1920s. (IIHR Archives)

the U.S. Army Corps of Engineers (COE) was assigned the duty of surveying the region's rivers and reporting on their possibilities for water-power, flood control, and navigation. Nagler accepted the Rock Island District Engineer's solicitation to assist with this effort for three years, several months of which he relinquished his university duties to work full-time at the Rock Island District office. His detailed reports of many of eastern Iowa's rivers, the first to be produced, set a standard for subsequent surveys performed by others and summarizing other drainages. Closer to home, he commenced long-term measurements of a small drainage near Iowa City (the Ralston Creek drainage), thus initiating what would become one of the most comprehensive, detailed data sets for a watershed now in existence.

In another effort, in 1929, he commenced work with the Iowa Board of Conservation (now Department of Natural Resources), which was then initiating a system of state parks and public lands. Nagler performed many of the site evaluations for future parks and constructed lakes, thus leaving a permanent mark on Iowa's natural and recreational landscape. He also became an active consultant for the *Iowa Twenty-Five Year Conservation Plan*, a landmark document that reshaped public management of the state's natural resources.

While performing river and park surveys, Nagler had ample time to exercise his historical leanings. He instructed his field crews to photograph and describe any remaining signs of nineteenth-century dams and mill sites. Today these slides of millsites remain our best remaining record of water-power during the settlement era. He also proceeded to bring rusting turbines, millstones, and related relicts back to the

University campus, where he shaped them into educational displays. Legend has it that with his determination and physical strength, he was able to lug millstones single-handedly from riversides, across fields, back to his Ford. His historical interest merged with a belief that Iowa's growing need for electricity could be met by waterpower from low-head dams, and he wrote several papers encouraging their construction.

Nagler's late-1920s river surveys led to additional assignments that provided the funds to carry the Hydraulics Laboratory through the depression-ridden 1930s. While working on the river surveys at the Rock Island District COE offices, Nagler entered discussions about plans to create cheap and reliable Upper Mississippi River transportation by the deepening river's navigation channel. Doing so would require the construction of dams to hold the water at the required 9-foot depth, and locks to allow the passage of boats and barges. Nagler reportedly was one of the first to recommend the 9-foot channelization project. He spent considerable time carrying out the necessary computations and advised the COE on the placement of the needed dams.

With Nagler's encouragement, when the dam at Hastings, Minnesota, was under construction in 1930, COE engineers were sent to Iowa City to work alongside Hydraulics Laboratory staff, who modeled the dam and a 20-mile stretch of the river. The results were sufficiently productive to establish the Hydraulics Laboratory as a major small-scale modeling site for federal projects in the eastern U.S. (Figure 3.).

Figure 3. Nagler-initiated COE model studies of Upper Mississippi River dams and spillways, performed at the Hydraulics Laboratory in the 1930s, carried the laboratory through the financial challenges of the Great Depression. (IIHR Archives)

The COE established a suboffice in the laboratory, whose research scientist ranks swelled with COE personnel. Soon small-scale models for several Upper Mississippi locks, dams, and spillways speckled the Hydraulics Laboratory. These COE model studies brought in the majority of revenue throughout the Great Depression, until the 1940s when war-related federal funding of basic research began in earnest.

Expanding a Laboratory, Creating an Institute

These diverse interests, energetically pursued, established a vigorous tone that affected the Hydraulics Laboratory's structure as well as its research and personnel. Already in 1923 Nagler had complained about the crowded conditions in the original laboratory. Additional space was sought in another Engineering College structure, but still Nagler bemoaned the fact that the Hydraulics Laboratory was turning away proposed projects because of overcrowded schedules and facilities. Five years later, in 1928, Nagler was taking charge of the construction of a three-story laboratory built on the same site as the old, but 60 by 30 feet in size, with experimental flumes, weighing tanks, measuring basins, and a circulating water system.

A few years later, in 1931, Nagler transformed the Hydraulics Laboratory into the Iowa Institute of Hydraulic Research, an organizational move that increased the lab's stature and ability to negotiate grants and contracts while simplifying business operations. And the following year – using the rental fees from governmental suboffices housed in the Hydraulics Laboratory – two more wings were added to the 1928 laboratory structure, forming essentially the building that remains today as IIHR's central administrative and office building, now an icon for hydraulics research and education around the world (Figure 4.).

With this structure, over 50 times larger than the original laboratory, Nagler provided IIHR's growing staff with a first-class laboratory for performing research, and with office and teaching space that was adequate for the times. These facilities provided a basis for future operations, remaining sufficient until 1948 when IIHR started to supplement its laboratory space with nearby research annexes. During a time when other laboratories were seriously challenged because of the nation's constricting economy, Nagler also had molded a firm financial base from the governmental suboffices. These suboffices fed a spiral of growing research efforts (the U.S. Geological Survey and Weather Bureau also moved into the new 1932 laboratory), and brought along personnel and projects that challenged IIHR's graduate students.

Leaving a Legacy

While overcoming considerable obstacles with his optimistic attitudes and creativity, Nagler was unsuccessful in meeting an internal challenge that presented itself late in 1933: a ruptured appendix. He underwent two emergency surgeries, but without today's antibiotics he died within a few weeks. He left three young children. His wife selected a massive uncut granite boulder as his tombstone, to symbolize his love of nature and rocks. His obituaries praised his "rare combination of native capacity,

Figure 4. The Hydraulics Laboratory, now IIHR's administrative center, was built in 1932 by Nagler and remains his most visible legacy today. (IIHR Archives)

thorough education, superb physical strength, fine enthusiasm, and high social purposes." They listed his professional awards – how he had received ASCE's Collingwood Prize for Juniors twice (in 1919 and 1920), the Norman Medal (with Albion Davis) in 1931, and the J. James R. Croes Medal (with David Yarnell) in 1932. They described his ability to inspire enthusiasm and energy in those with whom he worked, his dedication to church and family, and most of all the pleasure he took in his work and play – his "faculty of making his technical knowledge contribute to the pleasure of his pastimes" and vice versa. These comments reflect the standards and traditions Nagler had established at IIHR, which continue to this day: traditions of high and pleasurable productivity, excellence and innovation in research, and service to hydraulics and to the many agencies that brought their questions in search of solutions.

Nagler's death at the young age of 41 was completely unexpected. In the 11 following years, IIHR went through a sequence of three directors (Sherman Woodward, Clement C. Williams, and Francis M. Dawson) and four administrative arrangements, culminating with the appointment of Hunter Rouse as director in 1944. During this hiatus, IIHR continued to produce meaningful research and well-trained students. Perhaps most amazingly, it survived the Great Depression. These successes were the legacies of Floyd Nagler, who had constructed an impressive functional structure to house diverse hydraulic research, and established firm connections with governmental agencies that provided financial stability and innovative projects.

Connections with the COE, USDA, and other governmental agencies were terminated after World War II by Hunter Rouse, whose inclinations pulled IIHR away from contract and applied work and toward the basic research then being funded by NSF, ONR, and other post-war federal granting agencies. Rouse, like Nagler, took a firm, creative hold of IIHR's leadership and single-handedly guided its research and training program. His prominence soon overshadowed Nagler's, whose efforts and legacy are little known today. However none of Rouse's IIHR efforts, nor those of his colleagues and later research scientists, would have been possible if not for IIHR's founding father Floyd Nagler, who provided IIHR with a physical and administrative structure and a firm financial base, all overshadowed by the sheltering umbrella of his vision, enthusiasm, and dedication.

References

Crane, Jacob L. 1933. *Iowa Twenty-Five Year Conservation Plan*. Prepared for the Board of Conservation and Fish and Game Commission, State of Iowa, Des Moines IA.

IIHR Archives. (1920-1933) Floyd Nagler archives, Box 1. Hydraulics Laboratory, University of Iowa, Iowa City IA, 52242.

Ibid. (undated) Photo Archives. Hydraulics Laboratory, University of Iowa, Iowa City IA, 52242.

Freeman, John R., Ed. (1929). *Hydraulic Laboratory Practice*. American Society of Mechanical Engineers, New York NY. 868 pp.

Mutel, C.F. (1998). *Flowing Through Time: A History of the Iowa Institute of Hydraulic Research*. IIHR, University of Iowa, Iowa City IA, 52242. 299 pp.

Historical Improvements in Groundwater-Pumping Equipment and Effects on Farming in the United States

Bob Kent[1] and Greg Hamer[2]

Abstract

During the late 1800s and early 1900s, improvements in groundwater-pumping equipment significantly changed the lives of western farmers. These improvements also contributed to a higher standard of living and growth of the agricultural economy of the western United States. Without groundwater irrigation, agriculture was at the mercy of the seasons. Using groundwater, farmers had less to fear from droughts and seasonal variations in surface water. Better groundwater-pumping equipment also enabled farmers to increase the amount of land they could farm, increasing their profits.

The invention of self-regulating windmills and other pumping machines revolutionized farming, especially in arid areas of the Western United States. The wind-powered devices, which pumped at limited and uneven rates, eventually gave way to pumps powered by carbon-based fuels.

In the late 1800s, rapid improvements in mechanical devices led to more efficient oil- and gasoline-powered equipment that could lift and distribute ever-greater quantities of groundwater from greater depths to irrigate more acreage. An oil- or gasoline-powered pump could be left unattended for hours, enabling a farmer to oversee more acreage. The development of line-shaft turbine pumps and right-angle gear drives further advanced the use of deep groundwater for irrigation.

By the early 1940s, over 20 million acres of land in the United States was under irrigation. One result of improvements in groundwater pumping equipment was that large irrigated farms could take advantage of improvements in modernization and large scale agricultural practices.

---

[1] Vice President and Principal Hydrogeologist, Geomatrix Consultants, 330 West Bay Street, Suite 140, Costa Mesa, California 92627; phone (949) 642-0245; bkent@geomatrix.com.

[2] Senior Hydrogeologist, Geomatrix Consultants, 250 E. Rincon, Suite 204, Corona, California 92879; phone (909) 273-7400; ghamer@geomatrix.com.

Introduction

The use of groundwater for irrigated agriculture has increased significantly in the United States since the mid-1800s. Without groundwater irrigation, farmers were dependent on precipitation and precipitation-driven surface water supplies. Surface water supplies sometimes were limited by water rights and/or the price the farmer had to pay for water. In contrast, pumping from a private well is under direct control of the farmer, increasing his flexibility to irrigate outside the constraints of water rights or time of year. The increased use of groundwater for irrigation also enabled farmers to grow more crops in more areas. Improvements to pumping equipment and irrigating with groundwater enabled farmers to grow more crops with less labor and at lower overall cost, profoundly affecting their standard of living.

Before the mid-1800s, most farming in the United States was performed without the benefit of irrigation, relying primarily on precipitation. Except for large Southern plantations, farm size was limited by what the farmer and his family could cultivate. Between the mid-1800s and early 1900s, vast improvements in pumping equipment enabled farmers to lift and distribute more water to crops. Advances in pumps, motors, and wells built upon each other, not only helping individual farmers, but also producing far-reaching social and economic impacts and leading to further settlement of the Western United States.

Advances in Pumping Equipment

The industrial revolution of the late 1800s and the advent of steam, gasoline, and electricity to produce power had a major impact on farming in the United States. The evolution of groundwater-pumping equipment during the 40 years from about 1880 to 1920 was especially significant. Prior to the 1880s, farmers could economically lift irrigation water only a few tens of feet, a depth constrained by the size of the equipment that could enter hand-dug wells or pits. By the late 1920s, pumps and motors were available that could economically lift groundwater several hundred feet. The following describes historical improvements in pumps and pump motors in the United States and highlights the resulting changes in farming and farm economics.

Pre-1880. Before the 1880s, farm size generally was limited to the acreage that could be cultivated by family members (usually one or two acres). If sufficient reliable winds occurred in an area, a farmer could invest in a windmill and increase the amount of land he could irrigate, perhaps to 4 or 5 acres. Figure 1 shows a windmill typical of the time, alongside a water storage pond. Because windmills depend on wind speed and timing, they required sufficient storage capacity to hold the water pumped at night, when it was not used to irrigate fields. Water storage often consisted of an earthen pond or reservoir rather than a higher-cost steel or wooden tank. Before the invention of the self-governing windmill in 1854 (USDA, 2002), windmills, difficult to maintain under the best circumstances, proved subject to failure in high winds. Deeper wells became more prevalent in the 1870s, when drilling techniques became more widely available. Deeper drilling increased access to groundwater for irrigation and other uses.

Figure 1. Typical windmill and storage pond for irrigation (Haworth, 1897).

1880 to 1900. In the late 1800s, several innovations became available to lift water from wells. The Boyce water lift, shown in Figure 2, utilized horses to rotate a mechanism to repeatedly lift 50-gallon drums of water from a large-diameter dug well. In some cases the Boyce lift was used to extract water from multiple wells simultaneously. Other mechanical methods of irrigation pumping from wells included windmills and steam- or gasoline-powered pumps. The power source was usually connected to the pump by a belt (leather or other material) or shaft and gear mechanism. The power source and pumps could be located near each other or some distance apart.

The choice of pumping method, including the pump type and motor type, depended on the availability and cost of the pump and an energy source, such as wind, coal, or gasoline, and other factors such as the size and slope of the area to be irrigated and the lift required. The combination of a pump and power or energy source was often referred to as a "pumping plant". Most pumping plants could lift water only a few tens of feet; a lift of 100 feet was considered remarkable at that time.

Table 1 presents information from an 1890 census report that included costs for various types of water supplies (Wilson, 1896). The costs are expressed in two ways: the "first cost" and the "annual water rental." The "first cost" represents primarily the initial capital outlay for pumping and water storage equipment. This cost includes the power source or engine, the pumping mechanism itself, and any water storage tanks or reservoirs required. In some instances the first cost also included the cost of the land. The first cost is expressed per acre, as each type of water supply source lends itself to larger or smaller applications. A small farmer, for example, might not have the financial resources to pay for equipment that had a high first cost. A larger farmer might have the financial resources to pay a higher first cost and, thus, the ability to efficiently irrigate more land.

Figure 2. Boyce water lift (Hood, 1898).

The "annual water rental" refers to the operations and maintenance costs to irrigate an acre of land. Surface water sources, for example, might include a cost for water rights, whereas groundwater sources involve no water-rights costs. Cost *information* presented in this paper is as it appeared in the original references and has not been adjusted for inflation over time.

Table 1. Costs for Irrigation Water, Circa 1890

Source	First Cost per Acre[1]		Annual Water Rental per Acre[2]	
	Range	Average	Range	Average
Windmills	N.M.[3]	$20.00	N.M.	"practically nil"
Steam Pumping Plants	$5.00 to $10.00	$7.50[4]	$1.50 to $3.00	$2.25[4]
Gasoline Pumps	N.M.	$30.00	N.M.	$1.25

Notes: 1) The first cost for water includes pumping and/or irrigation facilities.
2) The annual water rental is the annual cost for irrigating an acre of land.
3) N.M. = Not mentioned in reference text.
4) No average value was presented in the original data. For this paper the mean value for the range of costs noted was calculated.

Windmills. Windmills were used widely in the late 1800s. Efficiency of irrigation using a windmill depended on a number of factors, including wind velocity, water lift, mill size, and pump cylinder size. A farmer had to evaluate the efficiency of the windmill he needed, considering most importantly wind speed and diameter of the wheel. An average wind speed of about 6 miles per hour was necessary to drive a windmill. Most windmills were manufactured to regulate the speed, because most were connected directly to a reciprocating pump, and slower pumping was more efficient. There was a negligible increase in efficiency in windmills having wheels greater than 12 feet in diameter because of the increase in weight and cost. For example, in order to resist windstorms, a 4-horsepower windmill would have to be eight times as heavy as a 1-horsepower windmill (Perry, 1899).

In selecting a windmill, a farmer had to consider the "duty of water," a term no longer in common use. The duty of water was the number of acres that could be irrigated with a given amount of water, a figure that varied greatly depending on the type of crop. Every crop requires a certain amount of water for each irrigation and a certain number of irrigations per growing season. Crops may require from 0.24 to 0.6 feet of water per irrigation, and from 2 to 7 irrigations per single crop (True, 1904).

Before 1900, windmills generally were more economical than steam-powered pumping plants, which required greater maintenance and operating costs. Although the first cost for a windmill is higher per acre, its annual water rental is considerably less than that of a steam-powered pumping plant. The cost of a windmill ranged from $50 to $400, and an earthen storage tank $100 or more in 1896 (Wilson, 1896). The average per-acre first cost was $20.00, and the annual water rental probably was significantly less than a dollar per acre.

An extensive survey of windmill pumping plants in Kansas indicated that between 1892 and 1903 the first cost ranged from $20 to $75 per acre. The higher end for costs reflected windmills that were underutilized. These costs compare with an earlier first cost of $20 per acre, as shown in Table 1. The annual water rental ranged from about $1 to $6 per acre, usually between $2 and $3 per acre, slightly higher than the earlier "practically nil" value reported in Table 1. Nettleton (1893) provided an example in which a windmill and equipment costing $135 were used to irrigate 4 acres of vegetables. The maintenance costs for the season were less than a dollar. After selling the crop, the farmer realized a profit of about $100 per acre.

Improvements to windmills increased the amount of acreage that could be irrigated. Although one windmill typically supported a few acres of crops, as much as 20 acres per windmill were reported (Wilson 1896). Using a windmill to irrigate 15 to 20 acres enabled a farmer to live better and even support a family. The increased yields that accompanied windmill irrigation also contributed significantly to local development of arid areas. The additional vegetables could provide food to local merchants and town dwellers. In addition, increasing the harvest of crops such as alfalfa would increase the number of cows or horses that could be sustained through the winter.

Steam-Powered Pumping Plants. Steam-powered pumping plants were in widespread use in the late 1800s. They were dangerous to operate, however, and required regular attendance by an operator. A steam-powered plant could be left unattended only briefly. Because of the supervision and repairs required, windmills were often preferred to steam-powered plants. Wilson (1896) states, "It appears that on the average the economy of a windmill is at least one and one-half times that of a steam pump, while it has an additional economy over the latter because of the attendance and repairs demanded by the steam boiler."

Wilson (1896) presents information on several types of steam-powered plants that could raise water as high as 80 feet and pump as much as 24 acre-feet per day (approximately 5,400 gallons per minute). In the early 1890s a survey of steam-powered plants in the United States reported a cost of about $4.00 to raise an acre-foot of water 100 feet (Nettleton, 1893). Steam-powered plants were used to irrigate several tens or even a few hundred acres. In one instance, a steam-powered plant in Eureka, Kansas, reportedly was used to irrigate about 3,000 acres (Wilson, 1896).

The capital cost for a steam-powered plant ranged from about $1,000 to $8,000, depending on the size and construction. An important economic aspect of steam-powered plants was the cost of fuel. Steam-powered plants used wood, coal, and later fuel oil. The average per-acre first cost was about $7.50. In describing systems used in the mid-1890s, Boyd (1897) mentioned several coal-powered pumping plants, including one 1,000-gallon-per-minute (gpm) system that consumed 1 ton of coal per day at a cost of $2.00 per ton. Elsewhere coal costs ranged from $1.50 to $2.00 per ton. In another example (Barker, 1898), fuel and operating costs for a 1,100-gpm steam-powered plant, accomplishing a lift of less than 30 feet, were $18 for each 30 hours of operation.

Figures 3a, b, and c illustrate a steam-powered plant that was used to lift water approximately 33 feet from a pit and well into an irrigation flume. Figure 3a is a diagram of the system, which included a steam engine and a rotary pump (Figure 3b) placed into a dug pit. The pump drew water from a 10-inch-diameter well, which was sealed to act like a suction pipe. The discharge from such a system is shown in Figure 3c. The large drive belt used to turn the rotary pump can be seen on the right-hand side of the photo. The belt is connected to a steam engine located to the right of the well.

Gasoline-Powered Pumping Plants. Gasoline pumps, a relatively new invention in the late 1800s, were first used to pump relatively small volumes of water, usually irrigating 5 to 50 acres (Wilson, 1896; Mead, 1904). Initially, the limited availability of gasoline limited the use of gasoline engines. With time, however, these engines, connected to various types of pumps, were used extensively in parts of the Western United States because they were less dependent on the local availability of fuel or water than were wood-, coal-, or water-powered pumps. Gasoline engines for pumps were relatively compact, simple to install and operate, and reliable. The largest such engines were capable of lifting 6 acre-feet per day (approximately 1,400 gpm) a height of approximately 20 feet (Wilson, 1896).

As might be expected, the average per-acre first cost for a gasoline pump ($30) was greater than for a windmill, as shown in Table 1. First costs for windmills and water or steam pumping plants were $20 or less. The overall cost of a gasoline plant ranged from about $400 to $600. Based on the information presented in Table 1, the annual water rental for a gasoline pump, at $1.07 per acre, was less than

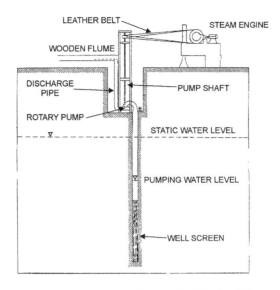

Figure 3a. Pumping plant near Las Cruces, New Mexico (Slichter, 1905).

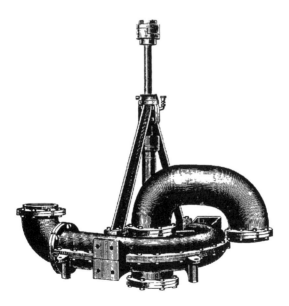

Figure 3b. Van Wie model rotary pump (Fuller, 1904).

Figure 3c. Irrigation well near Crowley, Louisiana (Fuller, 1904).

the cost for steam- or water-powered pumping plants. Gasoline at the time cost approximately 14 cents per gallon.

An evaluation of data from 175 steam- and gasoline-powered plants in several states (Mead, 1905) indicated that the cost to lift an acre-foot of water one foot vertically (an acre-foot-foot) varied by plant size, larger plants being more cost-effective. The data suggest that, on an acre-foot-foot basis, gasoline systems cost more than steam-powered plants. The cost per acre-foot-foot averaged $0.104 for gasoline, $0.045 for steam power from wood, $0.066 for steam from crude oil, and $0.048 for steam power from coal. Data from 70 plants indicated that the overall efficiency of the all pumping plants was only 44 percent. This was the ratio of the rated pumping plant horsepower to actual work performed in lifting water.

Pumps. Several types of pumps, requiring various power sources, were used commonly in the late 1800s. Centrifugal pumps, which came in a range of sizes, were used primarily for relatively low lifts of about 25 to 30 feet. A typical cost for a centrifugal pump was $100 per cubic-foot-per-second (cfs) capacity (per 450-gpm capacity). Rotary or revolving-piston pumps, capable of lifting water 100 feet or more, cost approximately $150 per cfs capacity. Mechanical elevators were also in use. These pumps, consisting of an endless chain or belt of buckets, were most efficient for low lifts of about 20 feet. A mechanical elevator capable of pumping as much as 5 cfs cost approximately $50 per cfs capacity (Wilson, 1896). Figure 4 shows how a centrifugal pump in Sacramento, California, was used to lift water from an 80-foot-deep well located at the bottom of a 20-foot-deep, wood-lined pit.

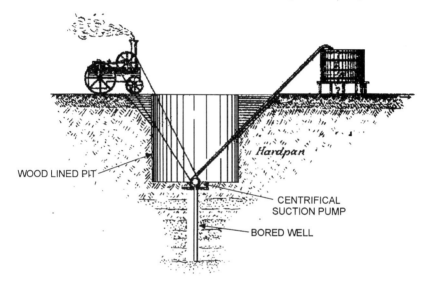

Figure 4. L.A. Oppenheim farm, Sacramento, California (Grunsky, 1893).

1900 to 1930. The early 1900s brought continued improvements in pumping equipment. In particular, wider use of deep-well centrifugal pumps, the invention of the right-angle pump drive, and the spread of electrical service brought significant improvements to irrigation efficiency.

Centrifugal pumps. The principal of centrifugal pumps has been known for more than 300 years (Driscoll, 1986). Until the advent of steam and other reliable power sources, however, centrifugal pumps were not widely used. In the late 1800s the popularity of centrifugal pumps increased, and by the 1920s they were widely used. For lifts of 25 feet or less, horizontal centrifugal pumps were often used. For lifts of 20 to 75 feet, a single-stage centrifugal pump commonly was used. This pump could be placed in the bottom of an open pit and connected to an electric motor or other power source by vertical shafting. A two-stage centrifugal pump could be installed where a deep vertical shaft was used to access groundwater. The cost and difficulty in excavating a deep shaft often limited the depth (and water lift) to 150 feet or less. Deep-well turbine pumps having several stages came into use in the late 1910s and early 1920s. At that time, Byron Jackson Company, for example, was manufacturing deep-well turbines in California. An advantage of the deep-well turbine pump was that it could accommodate wells that encountered a fluctuating water table; whereas contemporary single-stage centrifugal pumps connected to electric motors had difficulty where water tables fluctuated (Fortier, 1926).

A study performed from 1910 through 1912 of irrigated alfalfa farming in Southern California indicated that net profits ranged from about $26 to $39 per acre, depending on the year and the amount of water applied (Fortier, 1926). The market value of the crop is about $60 per acre. Considering that the crop requires about three acre-feet of water and that the profit averages $30 per acre, the cost of irrigation water is an important factor. If the cost per acre-foot of water increased by $10 and a crop required three acre-feet of water per acre, there would be no profit left, and the crop would be uneconomical.

Right-angle pump drive. The development of the right-angle pump drive had a significant impact on the ability of farmers, particularly those in the arid high plains, to pump groundwater for irrigation. Before the right-angle pump was invented, irrigation pumps were mounted in the bottom of pits hand-dug 30 or 40 feet deep. Maintaining the pumps and related equipment required dangerous descents into the pits. Additionally, the depth of the groundwater that could be accessed was limited by the maximum depth that a pit could be dug.

In 1902, Mahlon Layne invented the "pitless" pump (Reavis, 2002). Because this pump could work as far as 200 feet below ground surface, farmers in the high plains area could extract water from the underlying Ogallala Aquifer, which underlies a large area of the Central United States.

At that time, however, the pitless pump was still driven by motors connected to it by wide leather belts. The belts were inefficient, depriving the pumps of power. They went slack in the heat and taut in the cold, resulting in significant friction and loss of power. In 1914, however, the right-angle gear mechanism used in automobile differentials was adapted for use on irrigation pumps (Reavis, 2002). Although automobile motors were a natural choice for the new pump and right-angle drive mechanism, the motor speed was too fast, requiring additional modifications, which were quickly added to the design.

Electric motors. In the late 1920s, electric power was unavailable in most rural areas of the Western United States. As a result, many farmers relied on gasoline or other fuels to power pumps for irrigation. Fortier (1926) indicated that in localities where both cheap fuel oils and relatively cheap electric power were available, the cost for pumping water was slightly less using a gasoline engine. If fuel prices were high, however, electric motors had decided advantages, including lower repair and maintenance costs. As electric power became more widely available, electric motors became more widely used for irrigation pumps. Eventually, reductions in electric power costs made groundwater irrigation with electric-powered pumps more efficient and cost-effective than using gasoline or other fuels.

Changes in Farming

As described above, the nineteenth century brought a series of new developments in both pumps and the motors to drive them, enabling farmers to irrigate more efficiently and cost-effectively, thereby producing more with less effort. In 1830 about 250 to 300 labor hours were required to produce 5 acres (100 bushels) of wheat using a walking plow and other hand methods (U.S. Department of Agriculture [USDA], 2002). By the mid- to late-1840s, more advanced agricultural machines, irrigation, and the availability of commercial fertilizers had increased farming efficiency to the point that only 75 to 90 labor-hours were required to produce 2.5 acres (100 bushels) of corn. In the mid-1850s the design of the self-governing windmill had been improved, allowing wider use of windmills for pumping. In the 1860s, steam tractors became available, significantly increasing the amount of land a single farmer could cultivate.

The 1870s saw the first widespread use of deep well drilling, increasing the availability of water for irrigation. This change too required more powerful and efficient pumps to lift more water from deeper sources. By 1890, the number of labor-hours required to produce 2.5 acres of corn had declined to 35 to 40 hours; about half of what it had been 50 years earlier. The labor-hours required to produce 5 acres of wheat had declined to 45 to 50; less than one-fifth of that required in the 1830s. This decline in labor-hours per unit of production improved the life of farmers and enabled them to increase the amount of land they could farm.

Between 1900 and 1920 significant developments in pumping and irrigation equipment occurred. The right-angle pump drive was developed during this period, enabling farmers to more efficiently pump water from greater depths. Better pumps and power sources greatly increased the percentage of agricultural land irrigated using groundwater pumped from wells. Between 1909 and 1919, the number of acres in the United States irrigated by pumping wells almost tripled from approximately 500,000 to 1.3 million acres. During this time, the total amount of irrigated acreage in the United States increased more slowly, from about 14.4 million to 19.2 million (Fortier 1926).

By 1930, only 15 to 20 labor-hours were required to produce 2.5 acres of corn or 5 acres of wheat; about half the labor required in 1890. Overall, improvements in groundwater-pumping equipment and other farm practices had reduced the number of labor-hours required to grow corn and wheat by about 90 percent over the 100 years from the 1830s to the 1930s. Farmers were able to produce more than ever before. It is interesting, however, that over-production was often cited as a factor contributing to the decline in the number of family farms and the cause of the "great depression."

During the great depression of the 1930s, crop prices and land values fell more than 50 percent, forcing many farmers into bankruptcy. Many blamed the low prices on over-production, which resulted in passage of the Agricultural Adjustment Act of 1933. This act included price support and production controls that have had significant effects on farm economics. Since that time improvements in technology are not as important to farming economics as are price supports, the World Trade Organization, and the North American Free Trade Agreement, which together often receive blame for the decline of family farms.

By the 1980s further improvements in pumping equipment and farming practices had reduced the number of labor-hours required to produce corn and wheat to about 1 to 2 percent of that required in the 1830s. As groundwater-pumping equipment became more and more advanced and efficient, larger and larger percentages of the irrigated land in the United States were watered by pumped groundwater. Land could be farmed that otherwise would not be economically viable or practical to farm. Additionally, pumping of groundwater for irrigation provided for a longer growing season in many areas, because the water supply was not dependent on seasonal precipitation or the availability of surface waters.

Improvements in pumping equipment in the 100 years between 1830 and 1930 played a significant role in the development of agriculture in the western United States. Wider access to groundwater for irrigation and efficient, reliable irrigation equipment allowed farmers to put more land was into production and to produce more crops with less labor. These improvements gave farmers a higher standard of living, benefiting them, their families, and the nation as a whole.

References

Barker, F. C. (1898). "Irrigation in Mesilla Valley, New Mexico." *Water supply and irrigation papers, no. 10*, U.S. Geological Survey, Washington, D.C.

Boyd, D. (1897). "Irrigation near Greeley, Colorado." *Water supply and irrigation papers, no. 9*, U.S. Geological Survey, Washington, D.C.

Cowgill, E. B. (1897). "Irrigation practice on the Great Plains." *Water supply and irrigation papers, no. 5*, U.S. Geological Survey, Washington, D.C.

Darton, N. H. (1898). "Underground waters of a portion of southeastern Nebraska." *Water supply and irrigation papers, no. 12*, U.S. Geological Survey, Washington, D.C.

Davis, A. P. (1897). "Irrigation near Phoenix, Arizona." *Water supply and irrigation papers, no 2*, U.S. Geological Survey, Washington, D.C.

Driscoll, F. (1986). *Groundwater and wells*, Johnson Division, St. Paul, Minnesota.

Fortier, S. D. (1926). *Use of water in irrigation*, McGraw-Hill Book Company, Inc., New York, NY.

Fuller, M. L. (1904). "Underground waters of southern Louisiana." *Water supply and irrigation paper no. 101*, U.S. Geological Survey, Washington, D.C.

Grunsky, C. E. (1893). "Methods of applying water to land, as practiced in the central portions of California." In *A Report on irrigation and cultivation of the soil thereby*, Department of Agriculture, Washington, D.C.

Grunsky, C. E. (1898). "Irrigation near Bakersfield, California." *Water supply and irrigation papers, no. 17*, U.S. Geological Survey, Washington, D.C.

Haworth, E. (1897). "Underground waters of southwestern Kansas." *Water supply and irrigation papers, no. 6*, U.S. Geological Survey, Washington, D.C.

Hinton, R. J. (1893). *A Report on irrigation and the cultivation of the soil thereby*, U.S. Office of Irrigation Inquiry, Department of Agriculture, Part 1 of 4.

Hood, O. P. (1898). "New tests of certain pumps and water lifts used in irrigation." *Water supply and irrigation papers, no 14*, U.S. Geological Survey, Washington, D.C.

Hutson, W. F. (1898). "Irrigation systems in Texas." *Water supply and irrigation papers, no. 13*, U.S. Geological Survey, Washington, D.C.

Mead, E. (1905). *Annual Report of Irrigation and Drainage Investigations, 1904*, U.S. Department of Agriculture, Office of Experiment Stations, Bulletin 158.

Murphy, E. C. (1897). "Windmills for irrigation." *Water supply and irrigation papers, no 8*, U.S. Geological Survey Washington, D.C.

Nettleton, E. S. (1893). "Artesian and underflow investigation." *Final report of the Chief Engineer to the Secretary of Agriculture*, U.S. Department of Agriculture, Part 2 of 4.

Perry, T. O. (1899). "Experiments with windmills." *Water supply and irrigation papers, no. 20*, U.S. Geological Survey, Washington, D.C.

Reavis, D. J. (2002). "The right-angle pump." *Internet article presented at* www.texasmonthly.com/ranch/readme/rightangle.php, November.

Slichter C. S. (1905). "The rate of movement of underground waters." *Water supply and irrigation paper no. 140*, U.S. Geological Survey, Washington, D.C.

True, A. C. (1904). "Experiments, 1902-'04." Office of Experiment Stations, Fort Hays Branch Station, J.T. Willard, Director, *Kansas State Agricultural College Experiment Station, Bulletin 128*, December.

U.S. Department of Agriculture (2002). "A history of American agriculture, 1776-1990." *U.S. Department of Agriculture web site*: www.usda.gov/history2/tet4.htm.

Wilson, H. M. (1896). "Pumping water for irrigation." *Water supply and irrigation papers, no. 1*, U.S. Geological Survey, Washington, D.C.

It is a Crime To Design a Dam Without Considering Upward Pressure: Engineers and Uplift, 1890-1930

Donald C. Jackson[1]

Abstract

Mid-19th century gravity dam design did not include consideration of water pressure acting upwards on the base or within the structure of solid masonry or concrete dams. Technologies to relieve or reduce this upward press (usually termed "uplift") began to be implemented as early as the 1890s, but it was not until the failure of the Bayliss Paper Company's dam in Austin, Pennsylvania on September 30, 1911 that killed over 70 people that uplift engendered intense concern among American dam engineers. This paper will examine how the engineering community responded to uplift during the period following the Austin, PA disaster through the implementation of technologies such as foundation grouting, cut-off trenches, foundation drains, and drains within a structure, that were designed to obviate the destabilizing effect of uplift. It will also examine how engineers promoted buttress dam technology in the teens and 1920's, in part because widely spaced buttresses were not vulnerable to uplift pressures. By examining published books and technical articles as well as actual designs built in the period 1910-1930, it will be possible to document how engineers developed practical technologies to counteract the effects of uplift prior to the proliferation of dams during the New Deal era of the 1930s. Attention will be paid to the efforts of John R. Freeman (a president of the ASCE), Arthur Powell Davis (Chief Engineer and Director of the U.S. Reclamation Service), and others to both heighten professional awareness of safety issues associated with uplift and to develop designs that accommodated the destabilizing effect of uplift. This analysis of how engineers responded to uplift provides an important context for understanding the collapse of the St. Francis Dam in March 1928.

Introduction

In simple terms, "uplift" acts upon dams as the result of water in a reservoir percolating into the foundations (or into the dam itself), pushing upwards, and thereby reducing the ability of the structure to resist horizontal hydrostatic pressure. Uplift is particularly important in regard to gravity dam design because gravity dams rely solely upon the

---

[1]Associate Professor, History department, Lafayette College, Easton, PA 18042; phone 610-330-5171; inevery@yahoo.com

weight (or mass) of their structure to resist the hydrostatic pressure exerted by a reservoir. Uplift pushes upward in the opposite direction of the force exerted by the weight of the dam (which pushes downward). Thus, uplift acts to reduce the effective weight of the dam and reduce its ability to resist the horizontal pressure exerted by the reservoir. No dam foundation is ever entirely impervious to subsurface flow (just as no dam is entirely leak proof) and uplift is not something that, once present, will necessarily precipitate catastrophe. But it is a phenomenon that engineers should take into account when designing concrete gravity dams; otherwise, they risk disaster.

The deleterious effect of uplift can be countered through the technologies such as excavating cut-off trenches into the foundations, (which can reduce the ability of water to seep under the structure) grouting the foundations, (which fills in cracks in the rock foundation and helps block the flow of water under the structure) and by placing drainage pipes within the foundation and within the dam itself, (which allows water to be carried through tunnels and dispersed downstream from the dam). Uplift can also be countered by increasing the thickness of a gravity dam in order to increase the overall weight of the structure and thus offset the effect of uplift. As discussed below, all of these techniques were publicized and practiced by the early 1920's.

St. Francis Dam Failure

A few minutes before midnight on March 12, 1928, the St. Francis Dam gave way under the hydrostatic pressure of a full reservoir (Figure 1). During the early morning hours of March 13th, 38,000 acre-feet of water surged down from an elevation 1650 feet above the sea. Roiling through the Santa Clara Valley in Southern California, the flood wreaked havoc on several towns and scores of farms and rural communities. By the time it washed into the Pacific Ocean near Ventura at daybreak, more than 400 people lay dead.[1] Considered the greatest man-made disaster in modern California history, the failure of the concrete curved gravity St. Francis Dam brought renewed scrutiny to efforts by the City of Los Angeles (builder/owner of the dam) to expand its municipal water supply system and focused attention on William Mulholland, longtime head of the city's Bureau of Water Works and Supply and the official in charge of the dam's design and construction. It also brought attention to the issue of foundation conditions at dam sites and the existence of subsurface flow that could destabilize structures by pushing upwards under pressure and causing the phenomena know as uplift.

Immediately following the St. Francis collapse, several panels of engineers and geologists sponsored by the Governor of California, the Coroner's Jury, the Los Angeles City Council and others, prepared official reports on the disaster.[2] Although the panels were not in unanimous agreement on all points, within a few weeks a general consensus among these official panels espoused that poor rock beneath the western abutment softened when exposed to water seeping out of the reservoir. The softening of this rock was then alleged to have undermined the concrete gravity structure and caused the failure. After the west side blew out, water flowing across the canyon then supposedly began eroding the eastern abutment, precipitating a massive landslide that fed thousands of cubic yards of mud into the flood. In essence, weak foundations were generally held accountable for the structure's physical collapse, not the concrete gravity design, the construction methods used, or the specific effect of uplift acting on the dam.

In the aftermath of the disaster, dissenting voices soon appeared in the engineering press criticizing the almost lightening-quick judgment rendered by the various official engineering panels and pointing out the critical role that uplift had played in destabilizing the St. Francis Dam and precipitating its collapse.[3] But with the state legislature's passage of a new, comprehensive dam safety law in 1929 that placed the design, construction, and operation of all non-federal dams in California under the supervision of the State Engineer, (and with Mulholland's willingness to remove himself from public involvement in on-going projects related to Southern California's water supply) concern about the cause disaster largely faded from public attention.

In 1995, the *Southern California Quarterly* presented four articles related to the disaster that helped bring the tragedy back into the public limelight.[4] One of these articles by geologist David J. Rogers reconsidered the physical character of the dam collapse and reaffirmed that, in accord with objections raised in 1928 by engineers who disagreed with the conclusions reached by the official investigating panels, uplift acting against the base of the St. Francis Dam, and specifically against the east abutment, comprised the essential cause of the collapse. By combining fieldwork with a review of the historical record, Rogers offered useful insight into why the dam collapsed. But in

Figure 1. View of the St. Francis Dam in Northern Los Angeles County, CA shortly after its failure on March 12, 1928. This view is looking downstream and shows the center section of the dam that remained in placed. The view also shows how the flood scoured out the canyon below the dam. (D. C. Jackson, Historic Image Collection, reproduced with permission.)

acknowledging Mulholland's general lack of concern over the possible effect of uplift, he also promotes the impression that uplift represented a novel and little appreciated phenomena in the early 1920's when the St. Francis Dam was designed. (The design was prepared in 1923 and construction commenced in 1924.) Specifically he reports, "many engineers were just beginning to appreciate the destabilizing effects of uplift pressures in the late 1920s."[5] Had the dam been designed in 1910, such a statement might be justified, but uplift had been a topic of significant concern among American engineers in the decade before preparation of the St. Francis Dam design. The extent of this concern represents an issue deserving careful attention and comprises the focus of this paper.

Uplift history

Uplift had been acknowledged in the late 19th century and British engineers responsible for Liverpool's Vrynwy Dam, completed 1892, incorporated drainage wells into the design of this masonry gravity dam in order to counteract the effect of uplift.[6] This was publicized in British engineering journals in the mid-1890s, C. F. Courtney's 1897 *Masonry Dams: From Inception to Completion* noted that "the upthrust from the pressure of water tends to counterbalance the downthrust of the masonry, which may seriously affect the stability," while advising that "for securing the foundations against percolation a trench should be cut of not less than ten feet width at the top and three feet at the bottom, being of such a depth that as to completely cut out any false seam, spring, or vein of soft material, clay, etc.."[7] Uplift did not immediately attract the attention of civil engineers on a widespread basis. Some engineers such as Edward Wegmann, author of *The Design and Construction of Dams* first published in 1888 and in several subsequent editions, paid little heed, but this soon began to change. Notably, in the first edition (1910) of their book *High Masonry Dam Design*, the American engineers Charles E. Morrison and Orrin L. Brodie specifically observed:

> "In determining the cross-section by the series of equations developed in that analysis [by Wegmann for masonry gravity dam designs] no account is made of the condition of uplift due to water penetrating the mass of masonry, nor of the ice thrust acting horizontally... Present practice requires however, that these two factors be considered where a structure of great responsibility is proposed...."[8]

By 1910, concern about the effects of uplift had found voice in Morrison and Brodie's book, but it was not until the failure of the concrete gravity dam at Austin, PA. on September 30, 1911 that uplift attracted widespread attention among engineers in the American dam-building community (Figure 2). John R. Freeman, a prominent east coast engineer had worked with Mulholland as a member of the board of consulting engineers engaged to review plans for the Los Angeles Aqueduct in 1906. He visited the site of the Austin, PA tragedy, which killed more than 70 people, and wrote a letter/article published in *Engineering News* where he stated, "the cause that probably led to the failure of the Austin, PA dam... [was] the penetration of water-pressure into and underneath the mass of the dam, together with the secondary effect of lessening the stability of the dam against sliding."[9] He implored engineers to incorporate concern for uplift into their design procedures and advised, "uplift pressures may possibly occur under or within any masonry dam and should always be accounted for." Significantly, Freeman had recently served as a consulting engineer for New York City and helped design the Kensico Dam in Westchester County that, as noted in an April 1912 *Engineering News,* included consideration of "upward water pressure under the dam." The foundation for this 1,000 foot long, straight-crested gravity masonry dam was pressure grouted for its entire length and an extensive drainage system was also built along the length of the structure to help prevent seepage from causing uplift.[10]

Arthur. P. Davis, Chief Engineer (later Director) of the U.S. Reclamation Service had worked with Freeman and Mulholland in the fall of 1909 as a member of a board of consulting engineers for the Great Western Power Company. He also took the Austin PA failure very seriously.[11] Following a visit to the Austin site, in October 1911 Davis immediately wrote to Louis Hill, Supervisory Engineer for the Reclamation Service in regard to Elephant Butte Dam (a concrete gravity structure over 200 feet high to be built across the Rio Grande in southern New Mexico). He advised, "the failure of [the Austin Dam] was caused by an upward pressure on the base of the dam, due to the height of water in the reservoir."[12] The Reclamation Service subsequently approved a gravity design for Elephant Butte that was amply proportioned to accommodate the effect of uplift. Just as importantly, enormous care was taken to lessen the possible affects of uplift by undertaking extensive grouting of the entire foundation, placing an extensive drainage system in the foundations, and including a sizable cut-off wall at the upstream base of the dam (Figure 3). The foundation treatment of Elephant Butte Dam received

Figure 2. The Austin, PA. dam shortly after its failure on September 30, 1911. The view shows how the dam slid on its foundation and broke into large pieces as it moved downstream. (D. C. Jackson, Historic Image Collection, reproduced with permission.)

prominent public notice in *Engineering News, Engineering News-Record*, and in A. P. Davis' 1917 book *Irrigation Works Constructed by the United States Government*.[13] In the latter book Davis specifically stated, "a variety of precautions was adopted to prevent percolation under the dam, and to relieve any upward pressure that might develop there." The Reclamation Service's concern about uplift was hardly confined to Elephant Butte and Davis' book also describes the extensive grouting and drainage system implemented by the Service at the Arrowrock Dam in Idaho from 1913 to 1915.[14]

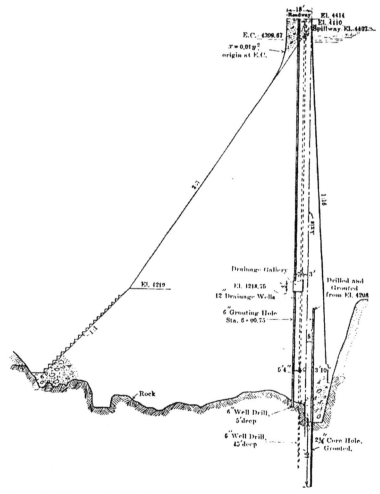

Figure 3. Cross-section drawing of Elephant Butte Dam in NM built by the U.S. Reclamation Service in 1913-1916. The drawing indicates the extensive grouting and drainage system placed in and under the dam. (Figure 79, Davis 1917)

In the aftermath of the Austin, PA failure, the perils of uplift were prominently featured and discussed in the engineering press. For example, C. L. Harrison's (1912) "Provision for Uplift and Icethrust in Masonry Dams," was published in the *Transactions of the American Society of Civil Engineers*. Taking up more than 80 pages of text, Harrison's article included discussion from 20 engineers who, while at times disagreeing on particulars related to the extent of uplift under various conditions, recognized that uplift represented a significant phenomenon warranting the attention of dam engineers.[15] The following year, *Engineering News* published a detailed report on recent field tests that measured, and confirmed the existence of, uplift pressures on dams in Germany.[16] This prompted feedback from engineers, and in a letter to *Engineering News* published on August 21, 1913, Edward Godfrey, who had taken part in the *ASCE Transaction* discussion on uplift the previous year, wrote that "the results of these experiments further emphasizes what the author has before said: It is a crime to design a dam without considering upward pressure."[17]

Along with coverage in professional journals, the importance of uplift as a design factor in the post-Austin Dam failure era also found expression in technical books. For example, Chester W. Smith's (1915) *The Construction of Masonry Dams* includes a 10-page section on uplift that discusses alleviating its effects through the use of cut-off trenches, foundation grouting, and construction of drainage systems.[18] The second edition of Morrison and Brodie (1916), now titled *Masonry Dam Design Including High Masonry Dams,* starts out with a 15-page discussion of uplift that provides the following guidance.

"There are several ways in which upward pressure may be cared for: First by adding a sufficient section to the dam to offset the upward pressure, and second by providing drainage wells and galleries to intercept all entering water, carrying it away through a discharge gallery or conduit to the lower side of the dam, and at the same time providing for as impervious an upstream face as possible. In the foundation an adequate cut-off [trench], of width and depth determined by examination of conditions disclosed during the progress of foundation excavation, is often advisable."[19]

Similarly, William Creager's (1917) *Masonry Dams* makes numerous references to uplift and incorporates it into several sample gravity designs. Creager takes care to specifically include uplift in the formulas used and explicitly includes consideration of the phenomena in its recommended design procedures.[20] Significantly, he explicitly incorporates uplift into the first of his "Designing Rules" that he describes in a chapter simply but forcefully titled "Requirements for Stability of Gravity Dams"

While Mulholland's design for St. Francis Dam did include ten drainage wells in the deepest section of the canyon, it should not be assumed that such limited use of drainage wells constituted standard contemporary practice. The Elephant Butte Dam discussed earlier utilized drainage wells along the length of the foundation for several hundred feet. In addition, The August 1916 *Engineering News* describes dams to be built in San Diego County featuring 12-inch diameter drainage wells spaced 12 feet apart, running the length of the structure. A cut-off trench also runs the entire length of the structure and a notation indicates a "continuous 12-[inch] sub drain [to be] laid along bottom of cut-off trench." The drawing also indicates extensive grouting along the length of the foundation and expansion joints spaced 96 feet apart.[21] Additionally, a 1921 *Engineering News-Record* article describes a 765-foot long, 135-foot high concrete gravity structure recently built in northern California where,

> "provision was made for grouting below the cut-off wall to make the seal at the upstream face as effective as possible. Furthermore, a network of underdrains was specified to carry off seepage water which might find its way under the structure... The drains under the dam consist of porous concrete tile cast on the job in 18-in sections, 3, 6, and 8 in. in diameter. Lines of were laid parallel with the axis of the dam and on 15-ft. centers under the entire structure."[22]

A 1924 *Engineering News-Record* article describes the 1,134-foot long, 184-foot high Black Canyon Dam in Southern Idaho recently completed by the Reclamation Service. This concrete gravity structure featured two rows of grout holes, each spaced at 10-foot intervals that were "drilled into the bedrock along the upstream edge of the dam along its entire length." The article further explained,

> "after the foundation had been grouted a row of drainage holes was drilled 8 feet downstream from the second row of grout holes... the water from them is carried to a tile drain embedded in the concrete parallel with the axis of the dam, discharging through the downstream face at frequent intervals. The purpose of this drainage system is to collect and lead of any water that might accumulate and to prevent an upward pressure under the dam."[23]

In 1919 construction began on a major curved gravity dam by the one other municipality in California that could compare in scale and scope with Los Angeles. Under the direction of Chief Engineer M. M. O'Shaughnessy, San Francisco built the Hetch Hetchy Aqueduct designed to bring water to the city from the far distant Sierra Nevada. A key component of this aqueduct system was the huge dam that would inundate Hetch Hetchy Valley in Yosemite National Park. The 430-foot high dam at Hetch Hetchy, named O'Shaughnessy Dam, is a curved, concrete gravity structure built with an extensive drainage system along the entire length of the structure and up both sides of the canyon walls. This drainage system utilized 1,600 specially designed and fabricated porous concrete blocks that were placed within the dam and the cut-off trench that extended the length of the foundation up both canyon walls. As *Engineering News-Record* described it in 1922, "the [3' x 3' x 3'] porous concrete blocks are placed in the bottom of the cut-off trench for its full length, and also in vertical tiers. The latter are 50 feet apart at streambed level and 12 feet apart from streambed level to the top of the dam."[24] O'Shaughnessy Dam's extensive drainage system and foundation cut-off trench unquestionably bore scant resemblance to what was done at St. Francis by Mulholland.

More examples of how engineers recognized, appreciated and accommodated uplift into the design of gravity designs prior to the late 1920s could be offered.[25] In closing this section, attention is directed to the fact that during the early 20th century some engineers were developing designs that specifically negated and obviated the problematic effect of uplift acting on gravity dams. Buttress dam designs that began to proliferate during the first two decades of the 20th century differed dramatically from solid gravity dams in that they did not present a solid, monolithic surface that subsurface water could push upwards on. In 1916, Salt Lake City specifically selected a buttress dam design for its 150-foot-high Mountain Dell because of concern for the quality of the foundation and the possibility that uplift could act upon a water storage structure built at the site. As Salt Lake City Engineer Sylvester Q. Cannon reported in a published article,

> "question[s] of stability and cost were carefully considered... After careful consideration of the various types, the multiple arch [buttress] type was accepted. One of the factors influencing this decision was the bedrock condition at the site. The bedrock is a calcerous shale not entirely watertight and of a nature to decompose somewhat under exposure to air and water. The advantages of the multiple arch [buttress] type in this connection were considered to be the practical elimination of upward pressure..."[26]

The above references document the concern about uplift that existed within America's dam-building community after the collapse of the Austin, Pennsylvania Dam in 1911 and prior to the design of the St. Francis Dam. And they demonstrate the actions that were taken by many engineers to counter the effect of uplift on the designs they were responsible for building. The point to be made is not that everything that could possibly be known about uplift was necessarily understood by American engineers in the mid-1920s. Certainly, dam engineers continued to focus on issues and problems related to uplift in the 1930s, 1940s and later through the 20th century. However, this subsequent work should not obscure our understanding and appreciation of the engineering insights related to uplift that were publicized and utilized in designs in the teens and 1920s. The point is that uplift was a phenomenon widely acknowledged and appreciated by engineers of the time. Furthermore, many engineers took significant action to counteract such destabilizing effects that extended far beyond what Mulholland did at St. Francis. In essence, Mulholland made a choice to do what he did at St. Francis, and this choice was at variance with professional practice as it developed and progressed in the decade after the failure of the Austin, PA. Dam in 1911.

Closing Comments

Perhaps the most remarkable aspect of the St. Francis Dam collapse does not relate to the issue of uplift *per se*. Rather, it involves the lightening quick completion of formal reports by panels of engineers and geologists that ascribe the collapse to a synergistic combination of "the human factor" and the "weak foundations" rather than to anything specifically related the technology of concrete gravity dams. Amazingly, the panel that reported to the Governor of California started work on March 19th and submitted their final report on March 24th, only six days later! Why the rush? Although never explicitly acknowledged in any records that this researcher has recovered, the desire to put the St. Francis Dam out of the public limelight as quickly as possible almost certainly relates to concern that it could derail Congressional approval of the Boulder Canyon Project Act which had reached a critical stage in the legislative process.

Boulder Dam (now known as Hoover Dam) utilized a concrete curved gravity design, which although quite different in its overall dimensions, nonetheless could be compared to the St. Francis Dam. Such comparisons needed to be squelched as quickly as possible if the desires of Californian politicians, developers and engineers who sought to tap the resources of the Colorado River were to be fulfilled. Had more time been taken to carefully assess and evaluate the failure of the St. Francis Dam, it is very likely that a "eastside/uplift" consensus could have been reached. This would however have required a public acknowledgment of the susceptibility of gravity dams to suffer from the effect of uplift and it could have fostered public concern perhaps, even unease, over the suitability of the Boulder Dam's gravity dam. The espousal of "weak foundations" and "the human factor" as reasons for the collapse of the St. Francis Dam allowed for a decoupling of the failed gravity design from the proposed Boulder Dam design it also helped quell the disquiet caused by the deaths of more than 400 people who were roused from their sleep in an horrendous and deadly crush of muddy water released from a reservoir created by the famous Los Angeles engineer William Mulholland. Rather than look to uplift as a supposedly new factor that offers useful insight into the history of the St. Francis Dam, I would propose that the economic and political context surrounding "the rush to judgment" in fact offers a more fruitful topic of contemplation for historians.

APPENDIX: End Notes

1. The best description of the disaster and its aftermath is provided in C. Outland (1963). *Man-Made Disaster: The Story of St. Francis Dam,* Glendale, California, Arthur H. Clark. Outland prepared a revised edition published in 1977 that provided additional details but made relatively few substantive changes to the original text.

2. For a good synopsis of these various official reports, see Nathan A. Bowers (1928). "St. Francis Dam Catastrophe - A Review Six Weeks After," *Engineering News-Record* 100, May 10, pp. 727-733. These include reports prepared by the Governor's committee, the District Attorney's committee, and the Los Angeles City Council committee. All of these committees' reports were completed and submitted in less than three weeks; the Governor's committee report was completed in six days.

3. C. E. and E. L. Grunsky (1928). "St Francis Dam Failure," *Western Construction News* 3, May 25, pp. 314-328; B. Willis (1928). "Report on the Geology of St. Francis Damsite," *Western Construction News* 3, June 25, pp. 409-413; H. P. Gillette (1928). "The Causes of the St. Francis Dam Disaster," *Water Works* 67, May, pp. 181-186; "Three Unreliable Reports on the St. Francis Dam Failure" (1928). *Water Works* 67, May, pp. 177-178; and C. H. Lee (1928). "Theories of the Cause and Sequence of Failure of the St. Francis Dam," *Western Construction News* 3, June 25, pp 403-408.

4. D. B. Nunis, Jr., and C. N. Johnson, eds. (1995). *The St. Francis Dam Disaster Revisited, Southern California Quarterly*, v. 77, nos. 3-4, Spring/Summer, Historical Society of Southern California/Ventura, CA, Ventura County Museum of History & Art.

5. D. J. Rogers (1995). "A Man, A Dam, and A Disaster: William Mulholland and the St. Francis Dam", in *The St. Francis Dam Disaster Revisited*, D. B. Nunis, Jr., and C. N. Johnson, eds., Los Angeles: Historical Society of Southern California/Ventura, CA: Ventura County Museum of History & Art., v. 77, nos. 3-4, Spring/Summer, pp. 1-110.

6. 19th and early 20th century theorizing by the German professors Lieckfeldt, Keil, and Link are noted in S. Leliavsky (1958). *Uplift in Gravity Dams*, Constable & Co., London, pp. 2-7. Also, I. Davidson (1987-88) "George Deacon (1843-1909) and the Vrynwy Works", *Transactions of the Newcomen Society* 59, pp. 81-95.

7. C. F. Courtney, (1897). *Masonry Dams: From Inception to Completion*, Crosby Lockwood and Son, London, pp. 51-52.

8. C. E. Morrison and O. L. Brodie (1910). *High Masonry Dam Design*, John Wiley, New York, pp. 1-3.

9. J. R. Freeman (1911). "Some Thoughts Suggested by the Recent Austin Dam Failures Regarding Text Books on Hydraulic Engineering and Dam Design in General", *Engineering News* 66, October 19, pp. 462-63.

10. A. D. Flinn (1912). "The New Kensico Dam", *Engineering News* 67, April 25, pp. 772-79.

11. The Great Western Power Company's board of consulting engineers that included Freeman, Davis and Mulholland is discussed in D. C. Jackson (1995). *Building the Ultimate Dam: John S. Eastwood and the Control of Water in the West*, University Press of Kansas, Lawrence, pp. 112, 284.

12. Arthur P. Davis to L. C. Hill, October 11, 1912; National Archives, Denver, Record Group 115, General Administration and Projects 1902-1919, Entry 3, Box 793, Rio Grande Project.

13. E. H. Baldwin (1915). "Excavation for Foundation of Elephant Butte Dam", *Engineering News* 73, January 14, pp. 49-55; E. H. Baldwin (1917). "Grouting the Foundation of the Elephant Butte Dam," 78 *Engineering News-Record*, June 8, pp. 625-628; and A. P. Davis (1917). *Irrigation Works Constructed by the United States Government*, pp. 243-45.

14. A. P. Davis (1917). *Irrigation Works Constructed by the United States Government*, New York, John Wiley, pp. 116-117.

15. C. L. Harrison (1912). "Provision for Uplift and Icethrust in Masonry Dams", *Transactions of the American Society of Civil Engineers* 65, pp. 142-225.

16. C. R. Weidner (1913). "Experiments on Uplift Pressure in Masonry Dams", *Engineering News* 70, July 31, pp. 202-205.

17. "Letter from Edward Godfrey" (1913). *Engineering News* 70 August 21, p. 371. In September 1913, *Engineering News* published an article on "Neglected First Principles of Masonry Dam Design". that called explicit attention to the relationship uplift between uplift and the need to increase thickness if (in the absence of effective grouting and drainage) stability was to be insured. A letter to the editor in response to this article trenchantly commented upon the financial implications of building safe, but bulky, gravity dams: "Only by the use of a wide base can there be safety in a gravity dam, and a wide base means a big bond issue". See G. H. Moore (1913). "Neglected First Principles of Masonry Dam Design". *Engineering News* 70, September 4, pp. 442-446; and J. S. Eastwood (1913). (letter regarding) "Neglected First Principles of Masonry Dam Design" *Engineering News* 70,October 23, pp. 32-33.

18. C. W. Smith (1915). *The Construction of Masonry Dams*, McGraw-Hill, New York, pp. 100-109.

19. C. E. Morrison and O. L. Brodie (1916). *Masonry Dam Design Including High Masonry Dams*, John Wiley, New York, pp. 8-9.

20. W. Creager (1917). *Masonry Dams*, John Wiley, New York.

21. "Arched Gravity Dams to be Built at Lower Otay and Barrett Dam Sites" (1916). *Engineering News* 74, August 12, pp. 195-96

22. "Concrete Dam on Eel River Built on Shale Foundation" (1921). *Engineering News-Record* 86, May 5, pp. 750-754.

23. "Building Black Canyon Irrigation Dam in Western Idaho" (1924). *Engineering News-Record* 93, November 20, pp. 818-823.

24. The porous block used for the drainage system at O'Shaughnessy Dam is described in "Plant and Program on the Hetch Hetchy Dam" (1922). *Engineering News-Record* 89, September 21, pp. 464-468. The extensive drainage system devised for the O'Shaughnessy Dam that extends completely up both canyon walls and through the entire dam is documented in the drawing titled "Drainage System Showing Galleries, Wells, and Contraction Joints - Sheet No. 3" signed by M.M. O'Shaughnessy in May 1919. A copy of this drawing is available in Folder 88, Charles Derleth Papers, Water Resources Center Archives, University of California, Berkeley, CA.

25. For example, see E. B. Whitman (1914). "The New Loch Raven Dam at Baltimore, Md.", *Engineering News* 72, August 13, pp. 331-337; "Exchequer Dam Construction Plant Built in Narrow Canyon" (1925). *Engineering News-Record* 94, May 28, pp. 880-884; "New Dam Will Double Water Supply of Portland, Ore." (1927). *Engineering News-Record* 98, May, 26, pp. 842-846; and "The Bull Run Storage Dam for Portland, Ore." (1929). *Engineering News-Record* 103, August 8, pp. 204-208. The latter two articles describe the 200-foot high, curved gravity Bull Run Dam, where construction commenced in May 1927, close to a year prior to the St. Francis disaster. Along with extensive grouting and an extensive drainage system, "at the upstream face a cut-off trench was excavated 6 feet below the finished foundation extending the entire length of the [900-foot long] dam".

26. S. Q. Cannon (1917). "The Mountain Dell Dam", *Monthly Journal of the Utah Society of Engineers,* 3, September, p. 226.

The Evolution of Pumping Systems Through the Early Renaissance

Paolo Macini[1] and Ezio Mesini[1]

Abstract

The hydraulic ability of the Roman engineers is well known, and is clearly shown in their design of aqueducts and open flow canals to deliver water to the main cities of the expanding empire. However, the hydraulic skills of ancient engineers were highly challenged when confronted with the problem of raising large volumes of water and no progress whatsoever was made in this field. Water pumping was a typical problem of the mining industry, where underground work was limited by the infiltration of water and aquifer systems. Consequently, the earliest application of heavy duty steam engines in the 18th century was to drive mechanical pumps to drain water from coal mines in England, marking the beginning of the so-called Industrial Revolution. However, before this revolution came about, in fact almost two centuries earlier, it is possible to trace back and discover the history of other mechanical inventions in the field of pumping. These are illustrated and described in technical treatises of the period and they led to the development of numerous types of pumping systems, some of which survived through to the end of the 19th century. The paper is aimed at giving a short outline of the development of pumping devices up to the early Renaissance, and highlighting the contribution of 16th century engineering to the design of mechanical inventions, which then paved the way to the industrial revolution of the centuries to come.

Introduction

Although the hydraulic skills of ancient civilizations were in some respects ingenious with regards to withdrawal, transport and distribution, they did not extend to resolving the problem of raising large volumes of water. The rudimentary pumps, already well-known in the Classical era and employed up to the late Middle Ages, were able to deal with modest flow rates and head capacity, and their power was very limited. As far as energy is concerned, antiquity developed only natural sources. Comparing the technological developments with the quantity of energy utilized, the Romans were at the same level of the Greeks. The main energy source was the muscular one, from draft animals and humans. Wind energy was utilized only for

---

[1] University of Bologna, Dept. of Chemical, Mining & Environmental Engineering, Viale del Risorgimento 2, Bologna 40136 Italy. paolo.macini@mail.ing.unibo.it; ezio.mesini@mail.ing.unibo.it

sailing, and the invention of the water mill did not revolutionized the technological application, as in the dry Mediterranean area rivers are often discontinuous water streams, their flow depending on the particular local climate. The low power of the machines available up until the end of the 15th century had also conditioned the development of the mining activity, and had seriously hampered the exploitation of numerous mineral deposits. With no adequate machinery for the draining of water from the mines, exploitation of the same was governed by the presence of an aquifer and its depth. Prior to the "modern era" of mining techniques (traditionally considered as being around the first half of the 18th century and marked by the joint application underground of gunpowder and steam engines), water drainage technology had already begun to progress in the first half of the 1500's, thanks to some remarkable discoveries and applications. Even the ancient mineral Codes (13th century) foresaw that there were serious limitations to the depth to which one could go underground, and they established precise regulations for the abandonment of underground excavations *(resignatio)*, with an aim to preventing the possible flooding of neighboring mines. However in the 16th century, the increased demand for raw materials and the exhaustion of the more superficial mineral deposits meant that mining had to go deeper everywhere. In those mining areas where it was indispensable to have permanent control over the infiltrations that invaded the galleries, remarkable progress was made in pump technology especially when the mining had to reach below the aquifer. Therefore, the evolution of pumping systems is closely linked to the development of mining techniques. Georgius Agricola's masterpiece, *De Re Metallica*, the first complete treatise on mining and metallurgy published in 1556, gives a complete report on the state-of-the-art. A long section of *De Re Metallica* is dedicated to pumping systems. The famous plates that illustrate Book VI are among the finest of the whole work and they provide, together with the accurate description of the hydraulic machines then in use in various central European mines, a vivid and precise picture of the different water drainage methods in use at the time.

Classical Antiquity and the Middle Ages

While mining activities in the Mediterranean during the Classical era are the subject of the rare accounts passed down to us by both the ancient Greek authors and of artistic works on vases and in mosaic, little or nothing tells us of the procedures related to the techniques in use at the time. This is because the techniques were the patrimony of the τεχνιτης (artisan) and, in the mining and metallurgical field, of the χαλκευς and the σιδηρευς (coppersmith and ironsmith). In fact they were not free citizens and they are, therefore, widely ignored in ancient documents. It is, however, likely that small seepage that impeded underground mining was solved by the simple and discontinuous method of bailing out the water with the aid of rudimentary equipment such as the *shaduf* (swape), originating from pre-classical times and still in use in many regions of the East. The next phase in the mechanical water drainage technology concentrated on achieving a drainage capacity that could be continuous. This was obtained by the application of two Greek inventions, the gear and the screw. With the former it was possible to drive, by means of a horizontal wheel turned by slaves or animals, the vertical wheels that carried buckets or amphorae (norias). The screw, on the other hand, led to the Archimedean screw. Accordingly to Diodorus, the ancient Egyptians already knew of this invention, traditionally attributed to the Syracusan. The Greeks were acute observers of natural phenomena and although their approach to problems was often purely theoretical, they did go on to explore. A favorite pastime of many a Greek inventor or philosopher was to study machines capable of defying the force of gravity. It was the famous school of mechanics that established itself in Alexandria between the third and first centuries B.C. that was particularly dedicated to these studies. Among the scholars were Philo of Byzantium, Hero of Alexandria and, above all, Ctesibius of Athens whose name is traditionally associated with the invention of the pressing and suction piston pump.

Ctesibius (3rd century B.C.), of whom no written work has survived, is considered the discoverer of the compressibility of air and was the first scholar to treat air as a material. As such, air was treated in the same way as the other fluids, and it could therefore be moved by suitable machinery. Ctesibius is attributed with the invention of the water clock and with perfecting the υδραυλος, or water organ. His inventions have been handed down to us mainly by the work of Hero (1st century B.C.) and Philo (2nd-3rd century B.C.). It is interesting to note the piston pumps were most probably fitted with flapper valves. Ancient technicians were not capable of machining movable sealing valves, so they had to resort to impractical devices that limited the diffusion of this type of machine. However, the Greek scientists, more theorists than engineers, had only established the physical principles and the static and geometrical characteristics of these machines. In fact, they concentrated on resolving the problems put forward by technicians. They were not very interested in a true and proper construction of machines, and only rarely can their devices be considered as true inventions. It was, therefore, a question of constructing, applying and improving these machines, and attempting to use a power source different than animals or slaves. This could not be addressed until the arrival of a pragmatic and applicative mentality, such as Rome and its engineers.

In mining, it was in drainage that Roman engineers differed from their Greek counterparts. The most promising applications were both the diffusion throughout the Empire of the water wheel, invented long before, and the improvement of norias and Archimedean screws. The wheel was applied at an early date to the task of raising water to great heights. There are two forms, the noria (or Egyptian wheel) and the chain of buckets, which is frequently, though inexcusably, confused with the noria. The noria consists of a large wheel having a series of containers fastened inside of the circumference. They are partially filled with water as they pass through the stream and they begin to discharge before they have reached the top. Thus, the water cannot be raised to the full height of the diameter of the wheel. When water was to be raised to greater heights, the noria was modified by attaching the buckets to a belt that passed over drums at the top and bottom of the lift. Such an apparatus had no specific limits beyond the amount of power available. Philo of Byzantium shows two devices of this type for use in wells of 15 cubits (about 22 feet). On the other hand, the primary features of the simpler form of the water screw are based upon the text of Vitruvius. This device was used to raise water for irrigation and to keep mines free from water. The particular convenience of the apparatus is its adaptability to raising water in restricted spaces.

Meanwhile, the standard use of ropes, pulleys, blocks and buckets continued. These were all moved with the help of machines that operated thanks to the rotation provided by drums. In turn, these were driven using toothed wheels and trains of gear, by horizontal or vertical wheels turned by animals or slaves (*molae asinariae*). The norias of the Rio Tinto mines (Tharsis, in the South of Spain) had a diameter of 4.5 m and held 24 containers on the external circumference. These had a capacity of 85 liters of water a minute over a depth of approximately 3.5 m. Some wooden water wheels were found in these mines in perfectly preserved condition (the Venafro wheel at the Museum of Naples and the above-mentioned wheels of Tharsis). This archeological find is therefore confirmation of the descriptions written in Vitruvius' work, which was well known and studied by numerous authors of writings on machines in the 16th and 17th centuries. Marcus Vitruvius, just before and right at the beginning of the Common Era, wrote a treatise (*De Architectura*) that is the only work of technical literature from the Roman world that has survived. The work, besides dealing with architecture, discusses technical questions relating to devices and machines, along the same lines as Hero's works. Book X is specifically dedicated to devices for the lifting and moving of weights, various hydraulic machines, *e.g.*, Archimedes' screw, mills, and Ctesibius' pump. In the Renaissance, Daniele Barbaro's translation of Vitruvius' work was of paramount importance since it contained numerous illustrations that had been re-elaborated in the light of the technical knowledge of the period. Vitruvius described the water mill (known with its Greek name υδραλετης). A century later Plinius illustrated numerous water mills installed on Italian rivers.

In Europe during the centuries following the fall of the Roman Empire, the demand for minerals and metals did by no means cease. However, the major mining centers moved from the Mediterranean to the new and more prolific mines of central Europe. It is significant that the miners of Saxony were the greatest masters of mining techniques of medieval and renaissance Europe. They began the exploitation of the famous mines of Schemnitz, in the Carpazi (745), of Goslar, in the Harz (970), of Freiberg, in Saxony (1170) and of Joachimsthal, in Bohemia (1516). The importance of the geographical and natural environment was paramount. During the Middle Ages, the western world was no longer centered around the Mediterranean, but moved far to the North. Since all technology depends heavily on natural conditions (and in the past even more so than today), many practices and techniques had to be re-adapted or re-invented. Also, the geographical conditions hampered or blocked the development of some techniques and machines invented in classical antiquity. Climate is definitely a severe constraint. In Northern Europe, rivers are larger, more regular in their flow, and do not dry up in summertime, as is common in the Mediterranean basin. Thus, the water mill saw an increase in use only in the Middle Ages, albeit it had been known since late antiquity. Indeed the full utilization of hydraulic power is no doubt the most important innovation of this period.

However, for more than a millennium, mining techniques and water drainage systems did not radically change, at least with respect to the equipment and methods. The only systems used continued to be the manual bailing out of water, norias,

Archimedean screws, chain of buckets and rudimental piston pumps. However, important progress was made with research into more powerful driving systems and ones that would not use slaves. Of great importance was the introduction of windmills, dating back to the late Middle Ages, even though their application in mining at that time is not certain. Equally important were the diffusion and the improvement of both the water-wheel and the crank motion system, a mechanical element that was unknown to the Classical age in that engineers had never added a fly-wheel to work as an energy accumulator, so indispensable if dead points were to be overcome. The drainage of water from mines required reliable and strong driving systems since the pumps had to operate continuously. In fact, as long as it was a question of draining water from rivers for irrigation, it was sufficient to have horses because the operation did not last very long. However, for mines it was better to find a continuous energy source. This was found in river currents and the technique was there and then defined. As early as the 6th century, the existence of mills was mentioned in local laws. Both water mills and windmills were, therefore, a part of the landscape in the late Middle Ages. Windmills were the preferred, in that they could be used in winter when rivers froze. Although an ancient invention, the water wheel is medieval from the point of view of its effective diffusion. In the 14th century, three positions were used for the water wheel in relation to the river current: the undershot wheel, the overshot wheel and the horizontal or breast wheel. Some wheels were mounted on boats in order to exploit the currents in the middle of the river. It is not easy to understand whether the angle of the paddles was taken into consideration in order to increase efficiency because the drawings of the time are not sufficiently detailed. Experiments were undoubtedly made in order to improve machinery and evidence of this is to be found, for example, in the many sketches of machines by Leonardo da Vinci in his Codes. He examined the problems relating to millwheels, and in particular he studied the position of the wheel, the shape of the paddles and the angle of incidence of the water. Leonardo's studies enabled him to formulate the principle of the hydraulic turbine, which had admittedly been understood by Hero, but had then been largely forgotten.

The incorporation of the art of "machines" and of "mechanical sciences" into the official doctrinal scholarship marks the originality of late medieval thought. This art allows one to imagine and to invent so as to modify natural matter through *ad hoc* contrivances, according to a numerical computation that allows one to obtain the predicted useful results. The famous sentence by Leonardo da Vinci "Mechanics are mathematicians' Heaven" marks the fulfillment of this new idea.

The birth of new technologies and the pumps of *De Re Metallica*

The knowledge of machines described in classical treatises was handed down to the Western world thanks mainly to the works of Arabian authors. Several treatises by Hero, including *Pneumatica*, that put forth various problems relating to pumping, were translated into Arabic by the order of the caliph 'Abbasid Ahmad ibn Mu'tasim around the year 830. Some Arabian contributions were also original. In 1206, the scientist Al-Jazari wrote a treatise on watch making and mechanics, in which he discussed a piston pump raised by a rack that was driven by cogged sectors. The noria takes its name from Arabian engineers, who, when repairing broken-down water drainage machinery constructed by the Romans, gave it the name *na'urah*. Furthermore, the first treatises on machines in the Western world during the Middle Ages stemmed from Arabian experience. A notebook written around 1265 by Villard de Honnecourt, master builder of cathedrals, was the first example of an engineering treatise. In a book written by Francesco di Giorgio (15th century), hydraulic pumps were mentioned, including the structure of the cylinder and piston, and a variety of gears, cams, link-blocks, cranks, cogged sectors and racks. When a lever operated the piston, counterweights balanced the forces to enable a more regular functioning. These modern machines were characterized by a widespread utilization of the crank drive system. This system both transmits and modifies the motion, transforming it from circular into linear, and vice-versa. It seems that this system was not known in antiquity and in the Middle Ages. The earliest drawing of such a device dates to around 1405. Such a mechanism does carry numerous drawbacks. From a manufacturing standpoint, it was difficult to produce mobile assemblies, and increased friction could take up most of the available power. Similarly, before the introduction of the flywheel, it was difficult to get past the dead points of the connecting rod.

In the first half of the 15th century, the windmill was used in the Netherlands to drain off marshes and flooded fields. Towards the end of the century, the windmill was used as well to drive sawmills, thanks to the use of the crank drive. Almost a century before the publication of Agricola's treatise, two new machines were introduced onto the European mining scene: the *pater noster* pump (or "rag and chain", see below) just before 1430, and the chain of buckets connected to a water wheel around 1470. For the first time, water coming in underground works could efficiently be kept under control and works could be opened below the deepest entrance gallery. However, the limitations of these machines were well known. Agricola pointed out that the *pater noster* pumps reached a maximum depth of 80 m, while the chain of buckets, driven by a water-wheel, could even reach down to 150 m, but they were so slow that they were able to eliminate small infiltrations only.

Georgius Agricola's *De Re Metallica* is considered as one of the leading 16th century treatises on machines. Existing originally as military engineering treatises, books on machinery began to cover almost every technical sector in the 15th and 16th centuries. Besides the precursors such as Mariano Taccola (*De Machinis,* 1438), Francesco di Giorgio (various treatises on architecture, engineering and military art, 15th century) and Roberto Valturio (*De re Militari,* 1472), the most mature authors writing on machines were Jacques Besson (*Theatrum Instrumentorum et Machinarum*, 1569), who was considered the initiator of this new literary genre, Guidubaldo dal Monte (*Mechanicorum liber*, 1577) and Agostino Ramelli (*Le diverse et artificiose machine*, 1588).

De Re Metallica discussed *metal machines*, namely those machines used in underground mining and for ore processing. In book VI, there are meticulous descriptions of the "devices with which water is drawn out of shafts, because the abundance of the same prevents mining". In fact, it often happens that the presence of consistent water infiltrations impedes the regular course of mining activities. Where there is no drainage gallery, a suitable collecting well has to be dug, "which like a sewer, or a bog, has to be lower", and a machine has to be positioned here for drainage up to the level of the nearest drainage gallery. A lengthy section of this book was dedicated to the description of these machines that are classified, after the winches, as raising machines, used to "pull out heavy weights from the shafts". With regard to this, Agricola pointed out that a breakthrough had been made and described a new kind of pump destined to revolutionize the mining techniques employed up until then: "of those hauling machines, forms are diverse and varied: & some of these being made with great skill: and if I am not mistaken, ancient peoples had no knowledge of any. They have been invented in order that water may be drawn from the depths of the earth to which no gallery reach."

Agricola, following the usual rigorous taxonomical criteria typical of his work, began to describe various kinds of pumps, after having acknowledged that "water is either hoisted or pumped out of shafts", namely it has to be removed in a discontinuous way using buckets, goatskins or tubs aided by ordinary devices or in a continuous manner using chain of buckets, reciprocating piston pumps or *pater noster* pumps.

Chain of buckets pumps. Agricola described three types of chain of buckets (or chain of dippers) pumps: the first (Figure 1) is a simple chain of buckets manually operated by means of an elaborate train of gears, acting as a reducer. This machine, of limited power, is characterized by both reduced capacity and by a drive system with low mechanical performance. "It carries out but little water and is somewhat slow, like all the other machines that have a great number of drums". It was for these reasons and also because of their high cost that they were not widely used. A second machine differs from the first in that there is no reducer: the system used is simpler and provides increased performance and consequently greater capacity because larger buckets can be mounted. On the other hand, this design requires greater power which, in this case, is provided by two men climbing up the inside of a squirrel cage (see below, and Figure 7). This method of propulsion, known and documented throughout the Classical era, was also used for most of the Middle Ages. The third type presented was probably the most efficient since it was operated by a water wheel and it was, therefore, possible to mount large-capacity buckets. The general disadvantage of these pumps was explained by Agricola with the words, "the buckets break down easily and for this reason miners use these machines rarely".

Piston pumps. Considerable text is dedicated to the discussion of single-action piston pumps, derived from Ctesibius' hydraulic pump and already known in the classical era. Agricola analyzed seven configurations of these machines that were improved designs compared to those used in the Middle Ages. They had a more precise construction of the coupling, the seal between body and piston, and the valves. The first style (Figure 2) illustrated the simplest system of all. The pump is operated manually, reciprocating the rod without the aid of levers. It is interesting to note the numerous details of the leather gaskets between the body of the pump and the piston, and the mechanisms of the flapper valves. Two other versions of piston pumps employ simple stands or stands balanced by a counterweight.

Figure 1. Chain of buckets pump, manually operated by means of an elaborate train of gears, acting as a reducer; this machine is characterized by both reduced capacity and by a drive system with low mechanical performances. (G. Agricola, *De Re Metallica*, Basel, 1556)

Figure 2. Single-action piston pump, operated manually by reciprocating rod without the aid of levers. Note the details of rods, pipes and flapper valves in the foreground (G. Agricola, *De Re Metallica*, Basel, 1556)

A fourth kind of piston pump began the series of multiple pumps, illustrated in some of the most beautiful plates of *De Re Metallica*. These plates are rich in construction details and in detail blow-ups, making here its first appearance in technical literature. This fourth variety of machine (Figure 3) was made up by two parallel pumps coupled together, and horizontal crankshaft drove the pistons. The pistons were connected to the crankshaft by means of connecting rods. Already in this kind of machine, still driven manually, one begins to realize the pump body has a low resistance to burst, and is susceptible to breaking apart. In fact, it was made from a log that is carefully drilled and milled with special tools in the piston stroke area (Figure 2). With regard to this, Agricola's advice was that "since it frequently cracks open, it would be better to use lead, copper or brass to make the body". An additional system involved the use of three parallel pumps. For this particular layout, the transformation mechanism of rotational motion into reciprocating motion is the shaft with square cams, more common in Agricola's "metal machines" than the crank drive. Technical specifications were given for this device: "with this machine water can be drawn from a great depth, namely 24 feet". If the machine was driven by a water wheel (Figure 4) a depth of 100 feet could be reached (it is reasonable to suppose that the *foot* used by Agricola corresponds to about 30 cm).

The last type of device described is of enormous importance in the history of mining and pumping technology. It overcame the intrinsic limitations of all the previous devices in which the pumping depth and the capacity were limited by the flapper valves and the length of the pump body that would be feasible and compatible with the stress imposed. In the new system, the length of the pump body was independent from the depth to be drained. The basic concept was simple. Single-action piston pumps are arranged in sequence, "one pump is underneath the other". Given this arrangement, the pumping depth could be increased as necessary, since the water is emptied into a small basin at each stage and the water in this basin was then moved up to the next stage. These machines were only suitable for installation in vertical shafts and needed considerable power because of friction. For this reason they were usually driven by large water wheels.

Agricola, emphasizing the innovativeness of this "metal machine", introduced around 1540 in the Saxon-Bohemian mining district, clearly recognized its application potentials. He in fact called it "ingenious, durable and useful, and not even costly". The simpler version (Figure 5) was a machine built with three piston pumps in sequence, the rods of which were synchronically driven by a single water wheel. One can see that the transmission mechanism, composed of shaped rods and small connecting levers, was kinematically redundant, so its performance, at least in the initial versions, must have been very poor.

Pater noster pumps. The treatise section dealing with the last six machines for water drainage regarded the *pater noster* pumps ("machines that draw up water by means of spheres"). This apparatus was nearly identical to the Cornish rag and chain pump of the same period. These pumps were built using a sliding closed loop chain inside a hollow pipe (the pump body), made from a perforated trunk identical to the one used in the piston pumps. The chain was fitted with spheres made from leather stuffed with horsehair and with the same diameter as the inside of the pipe. It slid along on

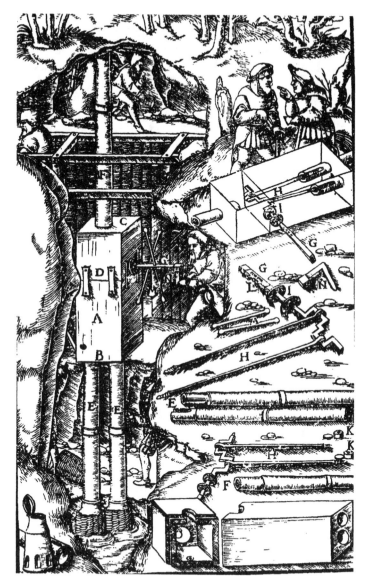

Figure 3. Piston pumps driven by a horizontal crankshaft, operated manually. Note the blow-up of crankshafts and connecting rods in the foreground (G. Agricola, *De Re Metallica*, Basel, 1556)

Figure 4. Three parallel piston pumps, driven by a shaft with square cams. (G. Agricola, *De Re Metallica*, Basel, 1556)

Figure 5. Three piston pumps synchronically driven by a water wheel. (G. Agricola, *De Re Metallica*, Basel, 1556)

the inside of the pump body by means of a friction pulley that was keyed into the axle of a water wheel. The leather spheres functioned like single-action pistons. The first type (Figure 6) was a machine moved by a large overshot water wheel and was fitted with "spheres where the hair of a horse's tail is sewn into the leather", spaced out at six feet apart along the chain. The attainable drainage depth declared for this machine was 210 feet with a 24-foot diameter wheel, and 240 feet when a 30-foot diameter wheel was used. A second type of machine was constructed with two pumps driven in parallel by two friction pulleys and the driving power necessary was, therefore, greater. A version driven by muscular power is also reported in Figure 7. The two automatic control systems associated with these machines are interesting. The first one was a level indicator for water to be drained. From a distance, the machine operator observed the position of this indicator (fixed to a float lowered into the shaft) and thereby knew when to activate the water wheel ("the worker opens or closes the canal in order to drive the machine"). The second device signaled the correct functioning of the system so as to ensure safety in the mine. This was a bell that rang as the chain of spheres moved. The bell was positioned on the surface and was set off by a mechanism at the bottom of the shaft triggered by the chain when it was moving. The presence of such a safety device may indicate that an operator did not always supervise these machines.

Another arrangement was the use of animal power, using elaborate horizontal wheels. Motion was transmitted to the pump by means of a wooden crown wheel and pinion, with the drive mechanism that was tried and tested in medieval mills. It seems that this pump succeeded in raising water from a depth of 240 feet using eight horses working in four-hour shifts. The treatise on pumps ends with a description of a large "metal machine". It was a device used for the raising of heavy weights (in this specific case large buckets and goat-skins for water). Its drum was keyed in axis with a 36-foot diameter water wheel that has a double set of paddles so as to be able to reverse the direction of rotation (Figure 8). A hoist operator ("rector of the machine") regulated, stopped and reversed the motion. The machine could be stopped by means of a braking cylinder, in axis with the water wheel. Five people were needed to operate the machine: the hoist operator, a brakeman, two workers for unloading and a worker at the bottom of the shaft for loading. Besides the remarkable dimensions of the whole machine, it is interesting to examine the mechanism devised for the reverse motion. This was based on the alternating entry of water onto paddles that move in different directions. However, this machine was only a powerful and versatile heavy-duty hoist. It did not feature any technical innovation with regard to the drainage of water.

A new machine for the drainage of water. The prospect that mines would be closed down on a large scale was an imminent danger as the underground works went deeper and deeper. This critical situation and lack of innovative technology was the boost that brought about the invention of a new machine that could resolve the problem of the restricted drainage depths provided for by machines up until the early years of the 1500s. This was accomplished by a series of piston pumps in sequence along the mine shaft, as described before. Such a machine, as described by Agricola (Figure 5) was still rather rudimentary and the absence of certain technical requirements meant

Figure 6. *Pater noster* (or rag and chain) pump, moved by a large overshot water wheel. (G. Agricola, *De Re Metallica*, Basel, 1556)

Figure 7. *Pater noster* pump, driven by the muscular power provided by two men in a squirrel cage. (G. Agricola, *De Re Metallica*, Basel, 1556)

Figure 8. Hoist used for the raising of large buckets and goatskins for water. (G. Agricola, *De Re Metallica*, Basel, 1556)

that it had to be replaced after just a few decades. In fact, the prototypes illustrated in *De Re Metallica* provided a rather small drainage depth, of around 20 m. Although the first machines were not well designed and extremely complex, they proved so satisfactory that they were immediately judged as durable, useful and, above all, economical. Their advantages were clear even to other mining technicians of the day who, just after a short time, recognized the enormous savings that were made possible by these systems. Accounts of the time tell of certain cases where the savings were around 90% of the total drainage costs. The successive applications of this kind of machine led to additional mechanical improvements. Just sixty years later, Georg Löhneiss (*Grundtlicher und aussfurlicher Bericht vom Bergwerk*, 1617) described a machine constructed with a set of 20 pumps that, despite it being complicated, was able to drain water at about 200 m. This demonstrates the impetus that existed in the second half of the century to update many pumping stations. The machine in question continued to have enormous success in all the European mines, showing both a wide geographical diffusion and long-lasting success in its applications. From Bohemia, it went to Saxony in around 1550, and from there it reached Harz during the following decade. By 1590, German miners had introduced it to central Sweden, in the famous mines of Kopparberg in Falun. In Southern Europe, the first machine was constructed in 1596 in the mercury mines of Idria, in Friuli. At this time the machine was also already known in Belgium, and probably also in Cumberland, England. The machine was well described in the treatise by Della Fratta (*Pratica Minerale*, 1678) as one of the most efficient pumps, and, thanks to the 18th century encyclopaedic works on machines by Andreas Bockler and Jacob Leupold, got into the *Enciclopédie* by Diderot and D'Alembert. These machines were further refined and improved. They were consequently linked to the driving force of large water wheels, and from the 18th century onwards, to the power of the new steam engines. In fact, the first industrial application of steam was made using this pump design. In England, Thomas Savery (*The miners' friend; or, An engine to raise water by fire, described*, 1702) patented his steam pump that was a rudimentary engine coupled with a multistage piston pump of the type described above. These pumps remained in widespread use until about 1870 when they began to be replaced in the more developed mining areas. In fact, these machines continue to be mentioned in some classic texts on mining until the first half of the 20th century.

APPENDIX: References and Supporting Literature

Clagett, M. (1959). *The science of mechanics in the Middle Ages*, U. of Wisconsin Press, Madison, WI.

Craddock, P. T. (1995). *Early metal mining and production*, Edinburgh U. Press, Edinburgh, UK.

Crombie, A. C. (1952). *Augustine to Galileo*, London.

Derry, T. K., and Williams, T. I. (1960). *A short history of technology: from the earliest times to A.D. 1900*, Oxford U. Press, New York.

Forbes, R. J. (1958). *Man the maker*, Schuman, New York.

Forbes, R. J. (1966). *Studies in ancient technology*, Brill, Leiden.

Gille, B. (1978). *Histoire des techniques*, Gallimard, Paris.

Kirby, R. S., Withington, S., Darling, A.B. and Kilgour, F.G. (1956). *Engineering in history*, McGraw-Hill, New York.

Koiré, A. (1961). *Les philosophes et la machine. Du mond de l'"a-peu-près" à l'univers de la précision*, Colin, Paris.

Maneglier, H. (1991). *Histoire de l'eau*, Bourin, Paris.

Marchis, V. (1994). *Storia delle macchine*, Laterza, Bari.

Prager, F. D. and Scaglia, G. (1972). *Mariano Taccola and his book de ingeneis*, The MIT Press, Cambridge, MA.

Russo, L. (1996). *La rivoluzione dimenticata. Il pensiero scientifico greco e la scienza moderna*, Feltrinelli, Milano.

Sarton, G. (1948). *An introduction to the history of science*, Harvard U. Press, Cambridge, MA.

Sarton, G. (1952). *Ancient science through the golden age of Greece*, Harvard U. Press, Cambridge, MA.

Singer, C. (1959). *A short history of scientific ideas*, Oxford U. Press, London.

Singer, C., Holmyard, E.J., Hall, A.R., Williams, T.I. (1957). *A history of technology*, Clarendon Press, Oxford, UK.

Thorndike, L. (1941). *A history of magic and experimental science*, Columbia U. Press, New York.

Tölle-Kastenbein, R. (1990). *Antike wassekultur*, Bech, München.

Usher, A.P. (1954). *A history of mechanical inventions*, Harvard U. Press, Cambridge, MA.

Hydraulic Engineering History at the Smithsonian Institution

William E. Worthington, Jr.[1]

Abstract

Over the last 250 years American hydraulic engineering has taken many forms. Some techniques, practices, and inventions were adapted from abroad. Others reflected an innate creativity. These sources are explored by discussing artifacts from the collections of the National Museum of American History. The focus of the paper will be on less well known events and people who made that history.

Introduction

The Smithsonian Institution is one of the few museums that includes hydraulic engineering as a major collecting and research area. Since the museum attempts to record and understand as much of our history as possible, hydraulics is defined in the broadest possible terms. This has resulted in a collection that encompasses a wide array of devices that use, operate by, or move water. Three-dimensional artifacts are enhanced by images, as well as documentation provided by written and printed records. Among the museum's hydraulic artifacts are pumps of all kinds, wooden pipe, waterwheels, Patent Office Models, testing and measuring instruments, valves, toilets, and material relating to canals, aqueducts, sewerage, and water treatment. While collecting possibilities are broad, it is obvious that not all hydraulics related artifacts are of a size or character that can be brought into a museum. In the case of objects that are too large, projects that are part of the landscape, or those which have disappeared altogether, historic photographs, prints, and paintings are invaluable substitutes.

While we can neither un-invent technology nor reverse events, it is possible to ignore both to such a degree that they appear to have never transpired. Hydraulics tends to go unnoticed by many, but it can be demonstrated that it is truly omnipresent and has been so throughout our history. However, many of the events were and still are local or regional and participants are not always widely known. But, taken together they add to our understanding of the past and they further illustrate that events do not occur in isolation.

---

[1] Specialist, Mechanical &Civil Engineering, National Museum of American History, Smithsonian Institution, Washington, DC, 20560; phone 202-357-1879; worthingtonw@nmah.si,edu

This paper focuses on historic images as artifacts of hydraulic engineering. It is a discussion of several less well known events that illustrate the breadth of hydraulic engineering. Yet, at the same time they demonstrate just how deeply integrated hydraulics has been in our overall development. They are typical of the interesting stories that can be uncovered when we look into the artifacts and records in the National Museum of American History.

Philadelphia

The first large-scale, steam-powered municipal waterworks in America was built because of the mosquito. Like other port cities, Philadelphia was plagued each summer by yellow fever. However, the siege of 1798 was different for it set in motion the plan for the formal adoption of a municipal waterworks.

Based on the belief that the fever was spread by impure water, Philadelphia's Common Council appointed a Water Committee to find a pure water supply. City residents took their water from local wells, but with an ever-increasing population, the quality of the water supply had deteriorated badly. Various plans were advanced that advocated bringing water into the city by way of canals from the rural areas to the west and north. Construction of one canal actually began, but progress was so painfully slow that the council turned to other means.

The committee ultimately adopted plans for the waterworks on March 2, 1799 and operations began on January 27, 1801. An engine located at the edge of the Schuylkill River pumped water to a nearby reservoir, from that point gravity carried it through a brick conduit to the Center Square pump house (Figure 1). A second engine located there filled cisterns in the upper story of the building. The cisterns in turn supplied water to the wooden mains that fanned out across the city.

The system was designed and built by English immigrant Benjamin H. Latrobe, who had knowledge of European waterworks. The centerpiece of his work was the grand marble edifice in Center Square. The building was classical in design and measured about 60 feet square and was noted for its domed cylindrical shape. Its familiar style disguised its function, for inside it housed not only the waterworks offices, but a Boulton and Watt atmospheric steam engine and its boiler. As the site was outside the developed part of the city, its pastoral setting became the destination for many outings. Its well tended grounds were enhanced by a fountain and wooden sculpture by Philadelphia artist William Rust. Not only did the site become a part of public consciousness, but in a benign way, so too did hydraulics as likenesses of the waterworks were used for decoration on ceramics and in widely distributed engravings.

Despite the immediate success of the waterworks, there were technical problems. Because of the limited reserve there was no water in the mains if either engine was out of commission for any length of time, and breakdowns were common. With two engines the problems only doubled. Decay of wooden components was an ongoing nuisance in both the engines and boilers. The overall inefficiency of the system was driven home in the many references to the huge amount of coal consumed by the boilers. The cost was simply too great and as a result, on September 7, 1815, the first municipal waterworks was taken out of service. A new efficient pumping station on the river at Fairmount was ready and set in motion.

Figure 1. Philadelphia's Center Square Waterworks as it appeared on a plate manufactured by Job & John Jackson of Burslem, England, c1831-1835.

The Center Square steam engine was sold and removed in 1818 and for a time the building served as a watchhouse for the western part of the city. Finally, in the late 1820s, the building was demolished. Today, the crossing of Broad and Market Streets where the waterworks once stood is the site of Philadelphia's City Hall.

Montgomery C. Meigs

Unlike their fellow citizens in Philadelphia, residents of Washington, D.C., took their water from wells, local creeks, and streams until the 1850s. In 1852, work began on a major federal government-sponsored engineering project that would supply the city with water diverted from the Great Falls of the Potomac River fourteen miles away.

It was a unique undertaking in the history of American cities for the project was designed and carried out by the U.S. Army Corps of Engineers. The Chief Engineer of the work was Montgomery C. Meigs. Initially he was appointed to survey the proposed route, but his work so impressed the Secretary of War that he was put in charge of the entire project. Over the next eight years a diversion dam was constructed, masonry and iron bridges were designed and erected, reservoirs built, miles of brick and stone conduits constructed, and mains of iron pipe installed. The largest and most visible part of the system was the stone arch conduit bridge over Cabin John Creek; for a time it was the largest single-span masonry arch in the world. Meigs cut a wide swath as an engineer. While he directed work on the aqueduct, he was in charge of the U.S. Capitol extension project that included the addition of east and west wings and the construction of the massive cast-iron dome. He was known for his ability to solve engineering problems creatively: the same arched iron pipes that served as structural members of highway bridges were part of the system that carried water to the city. And, there is strong evidence suggesting that his ego was as broad as his skills. By including his name wherever possible on his projects he insured that he would remain permanently associated with them. For example, open work in each riser of one set of cast-iron steps in the aqueduct project spells out M.C. Meigs.

It is no wonder, then, that the engineer when photographed with one of the systems's massive cast-iron valves included in the view his name and the name of the project on the side of the valve. Both names were part of the valve body casting and were in fact an added expense, as their presence required a modification of the wooden patterns used to make the part. The expense was hardly worth the effort--the valves would be seen by few for they were either buried directly in the ground or placed in underground vaults. In either case, their sides were hidden from view.

Meigs set up a machine to test the strength of the aqueduct system's cast-iron pipes. So determined was he to insure the success of his work, he sought to remove any defective parts before installation. Each length was subjected to a hydrostatic pressure of 21 kg/cm^2 (300 lb/in^2)--far in excess of the pressure it would experience after installation. After more than 150 years much of Meigs's system remains in use and, although hardly ever seen, his name remains largely in place.

Tide Mills

Nature has presented the engineer with a tantalizing number of powerful and seemingly inexhaustible forces which, when tamed, can be put to good use. The challenge has been to devise the means for harnessing them. For centuries wind has turned the sails of pumping and grinding mills and the flow of rivers and streams has turned waterwheels. But the wind doesn't blow all the time, nor does it blow everywhere and not all waterways have sufficient current or flow to turn a waterwheel. This is especially true in times of drought. These shortcomings led some creative minds to turn, with varying degrees of success, to the ceaseless movement of the waves and tides as unending sources of power.

The rhythmic rising and falling of the waves was seen as a possible alternative to the reciprocating motion found in some machines. Floats tethered at sea and riding the waves, could capture the motion, which might be put to use in place or transmitted ashore for use. In another type of motor, waves alternately filled and emptied a closed pipe, roughly approximating the action of a piston in a cylinder. A float located in the pipe moved up and down with the action of the water and through linkages, operated a pump. It should comes as no surprise that these motors had little impact on the field of energy production. Despite the potential of the various wave motors, they were at the mercy of the open sea and subject to its unpredictable and often devastating and destructive forces. Thus, there was little chance that any system based on wave movement could become both a permanent and reliable power source.

Quite apart from these methods, but nonetheless depending on the ceaseless motion of tidal waters, was the development of the tide mill. These mills, akin to the traditional waterwheel powered gristmills, were known to medieval Europeans and eventually spread across that continent. Examples were built in this country as early as the 18th century. In fact, a tidal mill at Poplar Grove Plantation near Matthews, Virginia, provided flour for George Washington's troops at Yorktown. As the operation of these mills was dependent on significant changes in the level of tidal waters, they were found only in coastal regions. Tidal mills were built from Maine to the Carolinas. Unaffected by any earthly force, including weather and temperature, the movement of tidal waters was as regular and predictable as the rising and setting of the moon.

In 1826 Caleb Hodgdon arrived in the coastal village of East Boothbay, Maine. He established a lumber mill and boatyard on a tidal pond at the mouth of the Damariscotta River. The pond covered about forty acres and only a narrow passage connected the two bodies of water. By damming up much of the opening Hodgdon was able to concentrate and control the flow of water into and out of the pond. His mill building was constructed on the dam and across the opening. The mill's operation depended on the difference in height of the impounded water and its source. At that location high tide was some 2.74 m (9 ft) above low tide. The power that could be extracted from water moving between the two levels was enough to drive the mill.

The operation of the mill was relatively simple and automatic. At high tide the force of the incoming water opened wooden gates positioned across the passage. When the level of the river began to drop, the gates closed. As the tide ebbed and after the level outside the gates receded about three feet, the impounded water in the pond was released through Hodgdon's undershot waterwheel. The mill had a typical interior and power from the waterwheel was transmitted by gears, shafts, and belts.

One of the major drawbacks to this and any other tide mill was that it could be used only on the ebb tide that occurred every twelfth hour. Also, tides occurred at a different hour each day. Altogether, power could be extracted for the six hours of the ebb tide and perhaps two more of the incoming tide. While there is no record of the exact size of the wheel Hodgdon used or the power it produced, it is known to have driven a large reciprocating saw as well as a grist mill in its early years.

In 1870 the old, slow-turning inefficient wooden waterwheel was replaced by high-speed hydraulic turbines made of iron. The highly efficient turbines produced as much as 97 kW (130 hp). By that time the factory included a whole range of woodworking equipment and the increased power was a necessity. Despite its deficiencies, the mill operated for more than a century, although in its final years, power came from a diesel engine. The old mill was finally removed in the summer of 1959 during a road-widening project.

It is interesting to note that this was not the only tide-driven machinery used on the coast of Maine. In 1906, the Rockland Power Company experimented with a system that used the tide to power an air compressor. Operating just as Hodgdon's mill had, water was impounded in a basin slightly over one mile in area. When released, it turned a waterwheel-driven compressor. The air it produced was stored in tanks and was then piped to area industries, where it was used to power equipment such as rock drills and machine tools.

The creative and fanciful methods for the use of what appeared to be free mechanical energy waned as the widespread distribution of water generated electricity became a reality.

Niagara Falls

The *Street Railway Review* proclaimed "The Falls Harnessed" in its 1897 review of the new Adams hydroelectric generating station at Niagara Falls. Actually, the statement was a good example of editorial hyperbole, as the falls remained untouched by the project, although a modest amount of the river's tremendous volume was diverted for the project.

In 1885 Thomas Evershed, an engineer for the New York State canal system, was asked to devise a plan for the further development of the waterpower at Niagara Falls. A number of companies were already perched on the wall of the cataract taking advantage of the vast power available at the site. Evershed's recommendations were comprehensive, but also called for the preservation of the natural beauty of the setting. Instead of using waterpower directly to run machinery, his plan called for it to generate electricity. The electricity in turn would be transmitted over wires for use at distant sites. Plants or stations established solely for the production of electricity were a fairly recent innovation, and an undertaking of such scope was unprecedented.

The Cataract Construction Company was formed in 1890 to carry out the plan. The nature and type of equipment that would go into it was based on investigations of successful European hydroelectric stations. In fact, the Swiss engineering firm of Faesch & Piccard was awarded the contract to design massive waterwheels capable of producing 3728 kW (5,000 hp). At the time of their installation they were the most powerful waterwheels built.

The station itself was located above the falls on the Niagara River and in no way detracted from the setting. Generating equipment was housed in a modest granite building designed by noted architects Sanford White and Frederick MacMonnies. The design had an official stolid appearance and there was no hint of grimy industrial use about it. The building was located a short distance from the river and water was brought in by a diversionary canal dug specifically for the purpose. The waterwheels or turbines were installed at the bottom of a 45.7 m (150 ft) deep wheel pit blasted out of the granite beneath the site. The power of the spinning turbines came from the head or pressure created when water fell the long distance from the canal to the wheels below. The most expensive and difficult part of the project was the work required to create the tailrace tunnel needed to carry off the water after it had passed through the turbines. A passage 2286 m (7,500 ft) long and 6.4 m (21 ft) in diameter was excavated in order to return the water to the river below the falls near the site of John Roebling's 1855 suspension bridge.

The system was relatively trouble free, but occasional maintenance was necessary and a visit to the bottom of the wheel pit was needed to remove obstructions lodged in the turbine runners. Although trash racks were installed above the water intakes in order to prevent accumulation, it was still possible for small items to work their way through. Were rocks or anything hard to lodge between the blades of the runner and the close-fitting exit gates, the turbine could be ruined.

Over the years the Adams Station was enlarged and equipment upgraded and the plant continued to produce electricity until 1961. At that time, the plant was permanently closed and its water diverted to a new station. Adams Station No. 1 and its companion plant No. 2 were emptied, the buildings razed, and the wheel pits filled in. All that remains of this first major American power station are photographs and the entrance archway to plant No. 1. After being dismantled, moved, and reassembled, the arch now stands on nearby Goat Island as a gentle reminder of the first major hydroelectric installation in this country.

Burden Waterwheel

For forty-five years the monster Burden waterwheel (Figure 2) stood 18.6 m (61 ft) high and 6.4 m (21 ft) wide. It was the largest and most powerful waterwheel in the world, and it drove an entire ironworks in Troy, New York. The wheel was the invention of Henry Burden, a Scotsman who immigrated to the United States in 1819. Although he had little formal training, Burden was an extremely creative mechanic and inventor who received patents for a variety of agricultural and iron working machines. His most important invention, one that had a positive impact on the success of the Union Army during the Civil War, was a horseshoe making machine. That device took raw metal stock and turned it into a completed shoe in seconds, and could produce more than 3,000 shoes per hour.

Burden began as a superintendent for the Troy Iron and Nail Factory and in 1848 became its owner. The great waterwheel he devised to power the factory was not a technical breakthrough, however. He took existing technology and simply made his wheel larger and more powerful than any other. Several factors influenced the decision to build a single huge breast-type wheel. Hydraulic turbines were relatively new and neither could a single turbine nor a single steam engine produce the great amount of power used by the factory. With steam engines there was also a constant need for fuel. Although a waterwheel required a steady supply of water with a reasonable head or pressure behind it, the cost was negligible.

His first wheel was built in 1838 and it operated satisfactorily for a number of years. But, in 1851 he constructed a new more powerful wheel. It was constructed like a giant bicycle wheel with individual adjustable wrought iron spokes that joined the rim to the hub. Needless to say, a large reserve of water was needed to supply the 127.4 cu/m (4,500 cu/ft) per minute consumed by the behemoth. The local stream was dammed ten miles away in order to produce an adequate reservoir from which to supply it. A conduit located atop a nearby hillside brought water to the giant wheel and as it left the six foot deep buckets. It exited through a pit into the nearby Hudson River.

Figure 2. The remains of Burden's factory and his waterwheel after the site was abandoned, c1900.

Getting power from the wheel to the factory was relatively simple. Around the circumference of the rim was a segmental gear that drove a pinion and floor-mounted line shafting that ran throughout the factory. While it could produce almost 373 kW (500 hp), it typically produced somewhat less than 224 kW (300 hp) at a speed of about 2 revolutions per minute. To maintain the wheel's structural soundness, it was kept wet and was roofed over so that it was actually an integral part of the factory.

By the mid-1890s, the factory was moved closer to the Hudson and power came from a modern steam engine. The great wheel was abandoned. Its brick foundation eventually collapsed and in 1913 the 238 ton (metric) (234 ton [long]) wheel dropped to the ground. There was talk of restoring the wheel in the early 1930s, but it came to nothing and by the end of that decade little remained but its metal fittings and they were dynamited apart for scrap. So ended what many had called the Niagara of waterwheels.

Humphrey Gas Pump

One of the most unusual pumping engines ever devised resulted from the work of H. A. Humphrey during the first decade of the twentieth century. He worked for a gas company in England and in response to his employer's request for a device that would create another market for the company's product. He devised a liquid piston pump. The pump functioned as a four-stroke-cycle internal combustion engine and used water for the piston. The process took place in a large, water-filled, upright U shaped pipe that was capped at one end. The cap served as a cylinder head containing intake and exhaust valves and a source of ignition. When an explosive charge of gas acted against the surface of the water it caused the water to rush toward the open end of the pipe. As the initial movement of the water was greater than the length of the pipe, some of the water was discharged. Following the explosion water that remained in the pipe began to oscillate from one end to the other. With each oscillation, the water performed the exhaust, intake, compression, and explosion (power) cycles of the four-stroke engine. As more water entered the pipe and the process was repeated rapidly--at operating speed--the pump produced a continuous flow.

The Humphrey Gas Pump Company of Syracuse, New York bought the rights to build this unique and now nearly forgotten pump. Although there were a modest number of Humphrey pumps constructed worldwide, only two examples were built in the United States. The first was installed in 1914 in Del Rio, Texas, where it was used to pump irrigation water from the Rio Grande River. Fuel came from a gas plant built to supply the engine. Gas was generated from the controlled burning of Mesquite trees harvested from the surrounding area. The pump remained in service only until about 1930 and its major parts were finally scrapped in 1952. The second example was built by the Sun Shipbuilding and Dry Dock Company of Chester, Pennsylvania, in 1925. That unit was a two-stroke-cycle design and was used only for experimentation. In the late 1920s an investigation was made into how the device might be used as part of a pumped storage hydroelectric station, but the project evolved no further. If there were additional tests involving the Humphrey pump they were unremarkable and went unrecorded. Today, the only remaining record of the American effort in this interesting and overlooked development in hydraulic engineering consists of several photographs and a few drawings. The pump itself was demolished in 1955.

Conclusion

Although only a handful of applications were discussed, they serve to illustrate some of the links between events, individuals, technology, and hydraulic engineering. Historic photographs open the past to investigation like no other artifact. When they are the only extant record, they can serve as a starting point in understanding--beginning with what is seen and then establishing connections to what is unseen. In the end it may be possible to comprehend why certain choices were made and others were rejected, and why events unfolded as they did. They demonstrate that a thoughtful discussion of historic images--photographs, engravings, and prints--can be a valid means of further understanding the history of hydraulics.

APPENDIX: Additional Information

Blake, N. M. (1956). *Water for the Cities,* Syracuse University, Syracuse.

"A Century-Old Tide Mill." (1902). *Power,* Sep. 19, 460.

Dunlap, O. E. (1900). "New Wheel Pit." *Engineering News,* Apr. 5, 229.

"The Falls Harnessed." (1897). *Street Railway Review,* 600-668.

Herschel, C. (1893). "The Niagara Turbines." *Cassier's Magazine,* Mar., 387-398.

Humphrey, H. A. (1910). "Humphrey Pumps and Compressors." *The Manchester Association of Engineers,* Manchester.

"May Restore Country's Largest Water Wheel." (1931). *The Locomotive,* Jul., 208.

"New Wheel Pit." (1900). *Engineering News,* Apr. 5, 229.

"The Old Burden Water Wheel." (1911). *Power,* Jan. 3, 8-9.

Penniman, H. W. H. (1902). "The Santa Cruz Wave Motor." Scientific American, Jan. 4, 8.

"Power Portal Honors Niagara Power Pioneers." (1967). *Niagara Mohawk News,* 30 July-August, 15.

Pursell, C., Jr. (1969). *Early Stationary Steam Engines in America,* Smithsonian, Washington D.C., 40-45.

Reynolds, T. S. (1974). *Stronger Than a Hundred Men, A History of the Vertical Water Wheel,* Johns Hopkins, Baltimore.

Rumsey, B. (1995) *Hodgdon Shipbuilding and Mills,* Winnegance House, East Boothbay.

Sellers, C., (1898). "Some of the Mechanical Features of the Power Development at Niagara Falls." *Transactions of the American Society of Mechanical Engineers,* 19:839-880.

"Sketches of Distinguished Inventors, Mechanics, and Manufacturers, Henry Burden." (1871). *American Artisan,* Feb. 1, 71-72.

Sweeny, F. R. I. (1915). "The Burden Water-Wheel." *Transactions of the American Society of Civil Engineers*, 79:708-726.

Swift, T. (1903). "Letters From a Practical Man, The Last of a Giant Water Wheel." *American Machinist,* Jun. 4, 803-804.

Thomson, F. D. P. (1934). "The Humphrey Gas Pump." *Mechanical Engineering*, Jun., 337-340.

Trump, C. C. (1914). "First Large American-Built Humphrey Pump." *Power,* Dec. 1, 767-770.

Webber, W. O. (1906). "The Use of Tidal Power for Compressing Air at Rockland, Maine." *Engineering News*, Dec. 6, 585-586.

Willson, B. D. (1984). "Using Historical Research in the Design of Liquid Piston Pumps." *American Society of Mechanical Engineers*, 84-WA/HH-4.

A Historic look at the USDA-ARS Hydraulic Engineering Research Unit

Sherry L. Britton[1], Gregory J. Hanson[2], and Darrel M. Temple[3]

Abstract

Over the past 65 years, scientists at the USDA-ARS Hydraulic Engineering Research Unit (HERU) in Stillwater, Oklahoma have made great strides in the design and technological development of hydraulic structures and vegetated channels. From its inception, the laboratory has gained notoriety for its accomplishments in vegetated channel design and development of design criteria for many hydraulic structures including trash racks, low-drop grade-control structures, and riprap design for rock chutes and stilling basins. Thousands of these structures and over a half-million miles of grassed waterways based on design criteria developed by the research scientists at HERU have shaped the landscape of the world into what it is today. In addition, research at the laboratory has provided field engineers with new techniques in measuring flow and soil erodibility. Today, scientists at the laboratory remain dedicated in their research efforts of hydraulic structures, channels, and issues associated with the rehabilitation of aging watershed flood-control and multi-purpose reservoirs. The purpose of the research has been and still is to assist engineers in making sound engineering decisions in their designs.

Introduction

During the Dust Bowl era in the 1930's, conservation of our soil became a national focus. Drought and poor farming practices followed by years of flooding caused severe erosion across the U.S. It was during this time that conservationists began

---

[1] Research Hydraulic Engineer, E.I.T.; USDA-ARS Hydraulic Engineering Research Unit, 1301 N. Western, Stillwater, OK 74075; phone 405-624-4135 x222; fax 405-624-4136; sbritton@pswcrl.ars.usda.gov

[2] Research Hydraulic Engineer, Ph.D., P.E.; USDA-ARS Hydraulic Engineering Research Unit, 1301 N. Western, Stillwater, OK 74075; phone 405-624-4135 x224; fax 405-624-4136; ghanson@pswcrl.ars.usda.gov

[3] Laboratory Director and Research Hydraulic Engineer, P.E.; USDA-ARS Hydraulic Engineering Research Unit, 1301 N. Western, Stillwater, OK 74075; phone 405-624-4141 x231; fax 405-624-4136; dtemple@pswcrl.ars.usda.gov

Figure 1. Aerial View of Lake Carl Blackwell and the ARS Hydraulic Laboratory.

developing best management practices for soil conservation. One of those practices, vegetative waterways, led to the establishment of the USDA-Soil Conservation Service (SCS) Outdoor Hydraulic Laboratory near Spartanburg, South Carolina in 1936. The initial intent of the laboratory was to answer questions related to the proper design of grass-lined waterways. The laboratory was moved to its present location near Lake Carl Blackwell, seven miles west of Stillwater, Oklahoma in 1941 (Figure 1). Now known as the Hydraulic Engineering Research Unit (HERU) of the Plant Science and Water Conservation Research Laboratory (PSWCRL), it became part of the USDA-Agricultural Research Service (ARS) in 1953.

Under the direction of project supervisor William O. Ree, the early-day purpose of the laboratory was to study the hydraulics of grass-lined channels, including terrace outlet channels, farm reservoir emergency spillways, diversions, and meadow strips. Research later expanded to incorporate hydraulic structures, including trash guards for closed conduit spillway entrances, hood inlet pipe spillways, and box inlet drops. A sister ARS Research Unit, conducting research on hydraulic structures, was located at the Saint Anthony Falls Laboratory in Minneapolis, Minnesota. The function of this research unit was moved to Stillwater and became a part of that program in 1983. In addition to research related to hydraulic structures, significant advances were made in the area of watershed runoff measurement through the development of new flow measurement devices and calibration procedures.

Since the early years of the laboratory, scientists have worked in cooperation with the USDA-Natural Resources Conservation Service (NRCS) (formerly the SCS). As an example, this cooperation has led to the development of Sites, Water Resource Site Analysis software used in design of watershed dams and spillways. In more recent years, research has included work in embankment overtopping, concentrated flow erosion, headcut erosion, and riprap design for rock chutes and stilling basins. Under Ree and his successors' leadership, the laboratory has gained national and international recognition as a significant contributor of sound design criteria for soil and water conservation structures and channels. Throughout its history, HERU has provided research assistance to provide information on the performance and guidelines for the design of many

Figure 2. Early research on vegetated-lined channels.

hydraulic structures. The focus of this paper is to give a historic perspective on a few of the contributions of the laboratory in the area of vegetated channels and hydraulic structures. Recent and future research is also discussed briefly in this paper.

A Chronological Timeline

1930's – During the 1930's, the American dream was slowly turning into a nightmare. The Great Depression was underway with the crash of the stock market, leaving the U.S. economy in chaos. Due to extremely dry conditions and poor management practices, agricultural lands of the Midwest were no longer producing the food they once did. Families packed their belongings and headed west in search of a new life, while leaving their old life and a devastated land behind them. Drought was prominent in the Midwest. The wind from the Great Plains swirled up clouds of dust that eventually blanketed cities on the East Coast. The Dirty Thirties, as it became known, began to take its toll on the land and its people.

With the growing interest in conservation, much thought was given to whether grass waterways could be used for erosion control. This interest led to the establishment of the USDA-SCS Outdoor Hydraulic Laboratory near Spartanburg, South Carolina in 1936. Its early day purpose was to study the hydraulics of grass-lined channels and to provide information to assist field engineers with combating soil erosion.

1940's – By the 1940's, the Great Depression and the Dust Bowl Days were fading memories. Taking the place of the dust storms were thunderstorms, dumping torrential rains and causing periodic flooding. In 1941, the Outdoor Hydraulic Laboratory was transferred to its present location near Lake Carl Blackwell, approximately 7 miles west of Stillwater, Oklahoma. Setting on approximately 100 acres downstream of Lake Carl Blackwell Dam, the laboratory uses a network of channels and piping to convey water to the different testing facilities around the grounds (Figure 1). Water is supplied through a large gravity flow water supply drawn by siphons from Lake Carl Blackwell. The flow capacity from these siphons is 130 cfs. The laboratory utilizes both indoor and outdoor testing facilities, having 4 model buildings, a large outdoor flume, grass-lined channels, and outdoor reservoirs for conducting research.

By 1944, America saw World War II come to an end and the beginning of the baby boom that resulted in increase housing and other development needs. These events along with the periodic flooding made America's farming community aware of their responsibilities in caring for the soil and water. With the end of World War II, SCS efforts were allowed to resume on flood control surveys and investigations along watersheds to determine whether flood control treatments would benefit the areas inundated (Geiger, 1955). With the passage of the Flood Control Act of 1944 (Public Law 534 of the 78th Congress), the SCS was authorized to begin eleven watershed projects for flood control.

During this same time period, research at the Hydraulic Laboratory was in full force. Under the direction of project supervisor William O. Ree, the primary objective of the laboratory was to study the hydraulics of grass-lined channels (Figure 2). This research effort resulted in the 1946 publication of SCS-TP-61, the *Handbook of Channel Design for Soil and Water Conservation*, a document that has been used worldwide for the design of vegetated channels for more than 50 years. Research on vegetated channels provided permissible velocity criteria and n-VR curves, both of which are valuable tools for the design of vegetated waterways (Cox and Palmer, 1948; Ree, 1949). In 1947 and 1948, research in vegetated channels was continued by testing various cultural practices such as the species, cutting, freeze and thaw cycle, dormant, burning, and other maintenance effects. To date, over a half-million miles of grass waterways has painted the landscape around the world. The work on grass-lined channels led to the eventual recognition of the laboratory as a National Historical Site by the American Society of Agricultural Engineers in 1990.

1950's – Because of the flooding events of the 1940's and 1950's, the HERU expanded its research into hydraulic structures, yet studies on vegetated channels continued. Beginning in 1950, the laboratory began investigating hydraulic structures such as the prototype-scale chutes with a SAF type stilling basin (USDA-ARS, 1981). The results of the tests showed the stilling basin could function effectively as a grade control structure at a lower cost than alternate basins. Between 1951 and 1953, research began on full-sized pipe outlet spillways for flood detention reservoirs. These tests yielded new data on friction factors for hydraulic design of straight section and miter-cut bends in pipes (USDA-ARS, 1981).

In December 1953, the Hydraulic Laboratory found a home with the newly founded USDA-Agricultural Research Service (ARS) (USDA-ARS, 1981). By 1954, a four-year study began on the friction factors for large diversion terraces planted to row crops (Ree, 1958). The tests were conducted with variations in row spacing, crops, and row directions with respect to flow. This research provided guidance in the sizing of waterways needed for terrace or diversion ditches with row crops. Findings from the study showed that the n-VR design approach could be applied to vegetation in rows as well as to randomly spaced grasses, for which the method was initially devised (Ree, 1958). In 1955 and 1956, studies were also conducted to investigate the effect of small, long duration flows on waterways located downstream of field irrigation. The findings concluded that grasses such as bermudagrass and Reed canarygrass could withstand long duration flows with little harm (Ree, 1976). By the late 50's and early 60's, hydraulic studies began on newly emergent vegetation planted for temporary protection of waterways (USDA-ARS, 1981).

Figure 3. Walnut Gulch flume model.

During the 1950's, research was conducted on flow-measuring devices. One device researched at the lab, using scale models (Figure 3), was the Walnut Gulch supercritical flow measuring flume designed in 1957 for use in southern Arizona on infrequent, flashy, and heavy sediment-laden flows. Current methods at the time were impractical or impossible because the flow conditions deposited transported material in conventional weirs and flumes, affecting their calibration (Gwinn, 1964). Extensive research was conducted to develop a new flume design that 1) remained clean of debris and sediment during and after each flow event, 2) would produce a rating relationship independent of the approach channel, 3) have a unique rating curve, 4) be free of backwater effects, and 5) have satisfactory accuracy under the operating conditions imposed upon it (Gwinn, 1964). A prototype Walnut Gulch supercritical measuring flume constructed in the early 1960's is shown in Figure 4.

1960's – Flood control reservoirs store runoff water temporarily, and a principal spillway is typically used to maintain the normal pool elevation behind a dam embankment. The principal spillway allows water to be slowly released at a rate to prevent flooding downstream. Along with the water, flood control reservoirs also become depositories for sediment and trash, including branches, grass, leaves, and man-made products. Debris such as this becomes a problem because wind and water currents typically direct the trash towards the reservoir outlets; thus, causing the outlet to plug with accumulated debris and the hydraulic performance of the spillway to be compromised (Figure 5). Trash racks are necessary to prevent the plugging of closed conduit spillways. In the late 1950's and early 1960's, research began on trash guards for

Figure 4. Walnut Gulch prototype flume.

closed conduit spillway entrances. Tests were conducted on different riser designs and trash rack modifications. This work was published in USDA Technical Bulletin No. 1506 (Hebaus and Gwinn, 1975).

Along with the work on trash guards, research began on hood inlet pipe spillways in 1961. Laboratory tests were conducted on full-sized hood

Figure 5. Accumulation of debris on a trash rack.

inlet pipe spillways, including 12-inch and 24-inch diameter standard corrugated, 8-inch and 12-inch diameter helical corrugated, and 12-inch diameter smooth continuous pipe (Rice, 1967). The results of the study on a full-sized smooth pipe hood inlet confirmed the results from previous model studies conducted by Blaisdell and Donnelly (1958). Additionally, the study showed that the full-sized rough pipe spillways did not perform as predicted by model studies on the smooth pipes. The rough pipes required a higher head to fill, and downstream tailwater conditions influenced the priming of the pipe spillway (Rice, 1967).

1970's – In the 1970's, research on trash racks continued primarily because of observations made concerning the performance of open racks. Debris accumulation on open racks caused by the drawdown of the approaching surface water affected the hydraulic performance of the spillway (Figure 5) (Gwinn, 1976). To contend with this problem, M. M. Culp of the USDA-NRCS conceived a design for a structure, known as the stepped-baffled trash rack, that he felt would eliminate some of the debris buildup, but he needed research on the structure to help support his idea. Gwinn (1976) began generalized research on the stepped-baffled trash rack at the

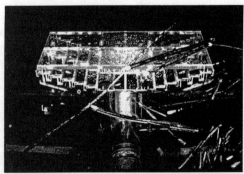

Figure 6. Model test of a stepped-baffled trash rack.

hydraulic lab (Figure 6). Research proved that the stepped-baffled trash rack is very effective in reducing surface currents that bring floating debris to the structure. The USDA-NRCS adopted this rack into a standard design, and it is one example of a trash rack that can be found in watershed reservoirs across the country (Figure 7).

In addition to the trash rack research, work was conducted during the 1970's to develop equations to determine head discharge relationships for any size H, HL, and HS flumes. Prior to this time, head-discharge relationships were determined by rating tables for flumes whose dimensions were assumed as ideal (Gwinn and Parsons, 1976). Further research on these flumes

Figure 7. Field application of stepped-baffled trash rack.

provided equations for determining head-discharge relationships for flumes whose dimensions were not ideal. Additional work was conducted on a drop box approach to the H, HL, and HS flumes (Figure 8). The development of these newly, designed flow-measuring flumes allowed bed load material to pass through the flume without a change in calibration (Gwinn, 1984).

1980's – In the early 1980's, research focused on the development of a stress based rather than the velocity based grass-lined channel design method (Temple et al., 1987). Research also evaluated the performance of vegetated earth channels used as auxiliary spillways on flood control dams and other hydraulic structures. Erosion of the spillway is considered permissible if the spillway does not breach during passage of the flood flow. Erosion is a complex process in earth spillways and is not completely understood. In 1983, the NRCS and ARS organized an Emergency Spillway Flow Study Task Group to observe and gather data from field spillways that had experienced greater than 0.9 m of head or had sustained major damage from flood flows. The goal of this group was to create a database and develop a better understanding of the erosion processes that occurred during vegetated earth spillway flows. HERU also initiated research in the laboratory on related erosion processes. This effort included additional study of the failure of grass channel linings and an improved procedure for the design of grass-lined channels, including grassed waterways (Temple et al., 1987). Studies were also initiated on surface and headcut erosion processes relating to spillway performance.

Figure 8. Drop box HS flume.

(Hanson, 1991; Robinson, 1992).

Additionally, studies were conducted in the late 1980's for riprap design for straight drop spillways and SAF stilling basins (Rice and Kadavy, 1991; Rice and Kadavy, 1992a; Rice and Kadavy, 1992b). Research showed excessive scouring downstream of straight drop spillways, placing the stability of the structure in jeopardy of failure. Tests conducted on physical models provided criteria for determining the size and placement of riprap downstream of the straight drop spillway to ensure the integrity of the structure (Rice and Kadavy, 1991). Rice and Kadavy (1992a) also examined the scour patterns upstream of straight drop spillways, and they found that the elevation of riprap relative to elevation of the weir crest, the average size of the riprap, flow rate, type of entrance abutment, and tailwater elevation all influenced the depth and extent of scour upstream of the spillway. The tests conducted resulted in the establishment of criteria to determine size and placement of riprap upstream of the straight drop spillway to ensure spillway integrity. In the end, this research provided better design tools to the SCS and a more economical design approach for sizing and placing riprap upstream and downstream of straight drop spillways (Rice and Kadavy, 1991; Rice and Kadavy, 1992a). By the late 1980's and early 1990's, research also developed design criteria for placing riprap in SAF stilling basins by using the Froude number and riprap size to determine the depth, width, and length of scour holes caused by impinging flows (Rice and Kadavy, 1992b).

1990's – By 1991, a new partnership was formed between the ARS and NRCS to address the complexity of erosion of earthen spillways. The team, known as the Design and Analysis of Earth Spillways, was charged with developing and documenting new technology for vegetated earth spillways. The team continued the research and field data acquisition efforts on vegetated channels and compiled the data for the development of a computational algorithm for use in design and analysis of earth spillways (NRCS, 1997). This computational algorithm was incorporated into the Water Resources Site Analysis computer program, Sites, used by NRCS for design and analysis of watershed flood control dams. The Hydraulic Laboratory, the NRCS, and Kansas State University continue to work together to maintain and upgrade the software as needed.

During the late 1980's and throughout the 1990's, research continued on headcut erosion processes, particularly research on gully headcut advancement (Robinson and Hanson, 1996; Hanson et. al, 1997; Hanson et. al, 2001; Robinson and Hanson, 2001). The development and movement of gullies in earthen auxiliary spillways is a dominant form of spillway damage. Because of the complexity of headcut erosion and the scale needed to perform headcut erosion tests, a large-scale flume was constructed on the laboratory grounds in 1993 for the primary purpose of conducting headcut erosion research (Figure 9). Throughout the 1990's, numerous tests were conducted in the flume with compacted cohesive soils of different geometry. Additionally, tests examined the variations in soil moisture content and dry density on headcut movement as well as variations in overfall height and flow rate. From the data collected over the past 10 years of research, Hanson, et. al (2001) has developed a simplified deterministic equation to predict headcut advance.

The continuing research on the erosion processes underlying vegetated earth spillway performance resulted in the development of a field and laboratory submerged-jet testing device to evaluate the erodibility of soils in auxiliary spillways (Figure 10) (Hanson, 1991). The development of this device led to a jet index characterizing the erosion resistance of a soil. The

Figure 9. Research on headcut advance in a flume.

device is a valuable tool for examining the effects of compaction effort and moisture content on erosion resistance of soil materials. To date, several patents have been awarded for the apparatus and procedure, and the procedure for determining the jet index appears as ASTM standard D5852-95 in the 2000 *Annual Book of ASTM Standards*. Since the development of this testing device, modifications have been made to characterize the soil erodibility of streambeds and stream bank materials (Hanson et al., 2002a).

Along with the earthen spillway research, generalized studies were conducted in the early 1990's on riprap design downstream of submerged pipe outlets (Rice and Kadavy, 1994a; Rice and Kadavy, 1994b). From this research, equations were developed to size the riprap required to protect the channel downstream of the outlet (Rice and Kadavy, 1994a). Relationships to size riprap-lined plunge pools were found to be functions of discharge, pipe diameter, tailwater elevation, channel bed elevation, and mean riprap (Rice and Kadavy, 1994b).

Additionally in the 1990's, generalized studies were conducted to develop criteria for the design of rock chutes on steep slopes. Rock chutes are loose-riprap-lined channels used to safely convey water from one elevation to a lower elevation (Figure11). These structures provide an alternative method for protecting the soil surface while maintaining a stable slope and are typically used for channel stabilization, grade control, and embankment overtopping protection. Previous design criteria for the NRCS did not exist for slopes greater than 10 percent, so studies were conducted at the laboratory to develop design guidelines for sizing riprap for slopes between 10 and 40% (Rice et

Figure 10. Laboratory submerged jet-test device.

al., 1998a; Rice et al., 1998b; Robinson et al., 1998a). Thousands of these structures have been constructed worldwide using the results of this research.

Other generalized studies conducted in the 1990's provided design guidance for low-drop grade-control structures (Rice and Kadavy, 1998c; Robinson and Kadavy, 1999). The results of these tests developed design criteria for determining the minimum height of the structure sidewalls, the need for wingwalls, and the size and placement of riprap for stability in the downstream channel (Rice and Kadavy, 1998c). Robinson and Kadavy (1999) took the research a step further by developing design criteria for a low-drop control structure with sloping sidewalls. Low drop structures have proven to be a viable solution for preventing additional degradation of the bed and banks of channels, and the research by Robinson and Kadavy (1999) provided alternative methods and materials for designing low-drop grade-control structures.

Figure 11. Prototype scale rock chute.

Known more for its generalized model studies, the laboratory has gained some notoriety in recent years for its specific model studies on roller compacted concrete (RCC) stepped spillways. Research for RCC stepped spillways for the states of Virginia, North Carolina, and Texas provided guidance on step configuration, wall convergence, and stilling basin design (Rice and Kadavy, 1996; Rice and Kadavy, 1997; Robinson et al., 1998b). Several of these structures have been designed and implemented as a result of these research studies.

2000's and the Future – Recently, research at the laboratory has become focused on challenges associated with the rehabilitation of earthen flood control and multi-purpose reservoirs. Many of the early NRCS reservoirs were constructed with a 50-year planned service life, and they're now reaching the end of this design period. Others have changed hazard classification due to downstream development or experienced changes in runoff due to changes in the upstream watershed. To continue to function as intended and to keep up with the forward progression and development downstream, these earthen dams are in need of rehabilitation. Rehabilitation may mean raising the top of the dam for increase storage capacity, widening the auxiliary spillway to obtain more flow capacity, and/or removing sediment stored behind the structure. Each of these rehabilitation methods can be costly, so another alternative being investigated to deal with these structures is the allowance of limited overtopping. Many questions are raised when addressing the overtopping of dams including: what acceptable depths of water are allowed over the embankment? how long is the water allowed to flow over the embankment? what effect does the embankment materials have on the allowable depth and duration of the flow? what rate does the embankment fail? and what rate of discharge occurs when the embankment does breach? To answer these questions, the Hydraulic Laboratory has initiated research related to the following topics: vegetal flow resistance and stress relationships on steep embankments, headcut migration through an embankment, and the widening of the embankment breach (Figure 12 and 13)

Figure 12. Analysis of hydraulic forces on vegetation in steep channels during overtopping.

(Temple and Hanson, 1998; Hanson et al., 2002b; and Hanson and Temple, 2002). Research at the laboratory is expected to continue on embankment overtopping issues in the future.

Summary

Hydraulic engineering research has progressed over the past 65 years at the USDA-ARS Hydraulic Engineering Research Unit in Stillwater, Oklahoma. From its first charter to investigate vegetated channels, research has expanded to include hydraulic structures that have shaped the landscape of the United States into what it is today. Research at the laboratory has assisted engineers worldwide in making sound engineering judgments as well as provided new techniques in measuring flow and soil erodibility. Under Ree and his successors' leadership, the laboratory has gained national and international recognition as a significant contributor of sound design criteria for soil and water conservation structures and channels. The Agricultural Research Service, of which the laboratory is a part, remains committed to providing U.S. agriculture with the technology required for sustainable production of a safe, high quality, food supply.

Figure 13. Headcut migration during embankment overtopping test.

References

Annual Book of ASTM Standards. (2000). D 5852-95. Standard test method for erodibility determination of soil in the field or in the laboratory by the jet index method. West Conshohocken, PA: ASTM.

Blaisdell, F. W. and C. A. Donnelly. (1958). Hydraulics of closed conduit spillways, part X, the hood inlet. Technical Paper No. 20, Series B, St. Anthony Falls Hydraulic Laboratory, University of Minnesota, Minneapolis, Minn.

Cox, M. B. and V. J. Palmer. (1948). Results of tests on vegetated waterways, and method of field application. Misc. Publ. No. MP-12, Oklahoma Agricultural Experiment Station, Oklahoma A and M College, Stillwater, OK in cooperation with the United States Department of Agriculture, Soil Conservation Service.

Geiger, R. L., Jr. (1955). A chronological history of the Soil Conservation Service and related events. United States Department of Agriculture, Soil Conservation Service (USDA-SCS), SCS-Cl-1.

Gwinn, W. R. (1964). Walnut Gulch supercritical measuring flume. *Transactions of the ASAE*, 7(3), 197-199.

Gwinn, W. R. (1976). Stepped baffled trash rack for drop inlets. *Transactions of the ASAE*, 19(1), 97-104, 107.

Gwinn, W. R. (1984). Chute entrances for HS, H, and HL flumes. *Jour. of Hydraulic Engr.*, ASCE, 110(5), 587-603.

Gwinn, W. R. and D. A. Parsons. (1976). Discharge equations for HS, H, and HL flumes. *Jour. of Hydraulic Engr.*, ASCE, 102(HY1), 73-88.

Hanson, G. J. (1991). Development of a jet index to characterize erosion resistance of soils in earthen spillways. *Transactions of the ASAE*, 34(5), 2015-2020.

Hanson, G. J., K. M. Robinson, and K. R. Cook. (1997). Headcut migration analysis of a compacted soil. *Transaction of the ASAE*, 40(2), 355-361.

Hanson, G. J., K. M. Robinson, and K. R. Cook. (2001). Prediction of headcut migration using a deterministic approach. *Transaction of the ASAE*, 44(3), 525-531.

Hanson, G. J. and D. M. Temple. (2002). Performance of bare-earth and vegetated steep channels under long-duration flows. *Transactions of ASAE*, 45(3), 695-701.

Hanson, G. J., A. Simon, and K. R. Cook. (2002a). Non-vertical jet testing of cohesive streambank materials. Paper No. 022119. American Society of Agricultural Engineers (ASAE) Annual Meeting, Chicago, IL.

Hanson, G. J., D. M. Temple, and K. R. Cook. (2002b). Research results of large-scale embankment overtopping breach tests. Association of State Dam Safety Officials Annual Conference, Tampa Bay, FL.

Hebaus, G. G. and W. R. Gwinn. (1975). A laboratory evaluation of trash racks for drop inlets. United States Department of Agriculture, Agricultural Research Service (USDA-ARS) Technical Bulletin No. 1506.

Natural Resources Conservation Service (NRCS). (1997) Earth Spillway Erosion Model. Chapter 51, Part 628, National Engineering Manual.

Ree, W. O. (1949). Hydraulic characteristics of vegetations for vegetated waterways. *Agricultural Engineering* 30, 184-187, 189.

Ree, W. O. (1958). Retardation coefficients for row crops in diversion terraces. *Transactions of ASAE*, 1(1), 78-80.

Ree, W. O. (1976). Effect of seepage flow on reed canarygrass and its ability to protect waterways. United States Department of Agriculture, Agricultural Research Service (USDA-ARS) Technical Report ARS-S-154.

Rice, C. E. (1967). Effect of pipe boundary on hood inlet performance. *Jour. of the Hydraulics Div.*, ASCE. 93(HY4), 149-167.

Rice, C. E. and K. C. Kadavy. (1991). Riprap design downstream of straight drop spillways. *Transactions of the ASAE*, 34(4), 1715-1725.

Rice, C. E. and K. C. Kadavy. (1992a). Rirprap design upstream of straight drop spillways. *Transactions of the ASAE*, 35(1), 113-119.

Rice, C. E. and K. C. Kadavy. (1992b). Riprap design for SAF stilling basins. *Transactions of the ASAE,* 35(6), 1817-1825.

Rice, C. E. and K. C. Kadavy. (1994a). Riprap design downstream of submerged pipe outlets. *Transactions of the ASAE,* 37(1), 85-94.

Rice, C. E. and K. C. Kadavy. (1994b). Plunge pool design at submerged pipe spillways. *Transactions of the ASAE,* 37(4), 1167-1173.

Rice, C. E. and K. C. Kadavy. (1996). Model study of a roller compacted concrete stepped spillway. *Jour. of Hydraulic Engineering* ASCE, 122(6), 292-297.

Rice, C. E. and K. C. Kadavy. (1997). Physical model study of the proposed spillway for Cedar Run sit 6, Fauquier County, Virginia. *Transactions of the ASAE,* 13(6), 723-729.

Rice, C. E. and K. C. Kadavy. (1998a). Roughness of loose rock riprap on steep slopes. *Jour. of Hydraulic Engineering,* ASCE. 124(2), 179-185.

Rice, C. E., K. C. Kadavy, K. M. Robinson, and K. R. Cook. (1998b). Rock chute outlet stability. *Transactions of the ASAE*, 14(2), 145-148.

Rice C. E. and K. C. Kadavy. (1998c). Low-drop grade-control structure. *Transactions of the ASAE,* 41(5), 1337-1343.

Robinson, K. M. (1992). Predicting stress and pressure at an overfall. *Transactions of the ASAE*, 35(2), 561-569.

Robinson, K. M. and G. J. Hanson. (1996). Gully headcut advance. *Transactions of the ASAE,* 39(1), 33-38.

Robinson, K. M., C. E. Rice, and K. C. Kadavy. (1998a). Design of rock chutes. *Transactions of ASAE*, 41(3), 621-626.

Robinson, K. M., C. E. Rice, K. C. Kadavy, and J. R. Talbot. (1998b). Energy losses on a roller compacted concrete stepped spillway. Water Resources Engineering 1998 ASCE Conference, Vol. 2. Memphis, TN. 1434-1439.

Robinson, K. M. and K. C. Kadavy. (1999). Hydraulic modeling of a low-drop grade control structure. Water Resources Engineering 1999 ASCE Conference, Seattle, WA.

Robinson, K. M. and G. J. Hanson. (2001). Headcut erosion research. Proceedings of the 7th Federal Interagency Sedimentation Conference, Reno, NV.

Soil Conservation Service (SCS). (1947). (Revised 1954). *Handbook of Channel Design for Soil and Water Conservation.* SCS-TP-61.

Temple, D. M., K. M. Robinson, R. M. Ahring, and A. G. Davis. (1987). Stability Design of Grass-Lined Open Channels. USDA, Agriculture Handbook No. 667. USDA, Washington, D. C.

Temple, D. M. and G. J. Hanson. (1998). Overtopping of grassed embankments. Proc. of the 1998 Annual Conference of Association of State Dam Safety Officials, Las Vegas, NV. (CD-ROM).

United States Department of Agriculture, Agricultural Research Service (USDA-ARS). (1981). Program planning and review of the Hydraulic Engineering Research Unit in Stillwater, OK on Jan. 15-16, 1981.

History of Hydraulics and Fluid Mechanics at Colorado State University

Pierre Y. Julien[1] and Robert N. Meroney[2]

Abstract

The historical contribution of Colorado State University to hydraulics and fluid mechanics is reviewed in reference to those who pioneered the analysis of hydraulic and wind engineering. The article first covers the "early developments" with Elwood Meade, Charles Lory, Ralph Parshall and Emory Lane during the Colorado A&M period. The name change to Colorado State University initiated considerable expansion through fruitful collaboration with the U.S. Geological Survey and the U.S. Bureau of Reclamation. The "expansion years" featured the contributions of Maury Albertson, Hunter Rouse, Jack Cermak and Everett Richardson under the leadership of Lionel Baldwin, Daryl Simons and Ray Chamberlain. The more recent "mature period" saw broadening of the programs and expansion into environmental engineering. Some of the key scientific achievements are reviewed, and the interaction between faculty activities, academic programs and research facilities that led to rapid growth and development are retraced. The success and visibility of the hydraulics and fluid mechanics programs also hinged on several other factors including significant contributions at the international level through projects in Pakistan, Columbia, and Egypt.

Introduction

In 1862, the United States Congress passed the Morrill Act granting land to each state in the amount of 30,000 acres for every senator and representative. As a result of this Act, the Territory of Colorado received 90,000 acres. The receipts from the sale of this land were designated for perpetual fund to support "at least one college to teach agriculture and the mechanic arts". Although the 1862 Act gave the state a grant of federal land for a college, it was not necessary to locate the college on that land. In 1870 Governor Edward M. McCook signed the Territorial Bill establishing the Agricultural College of Colorado at Fort Collins. In November of 1874 the first

---

[1] Borland Professor of Hydraulics, Department of Civil Engineering, Colorado State University, Fort Collins, CO 80523-1320, 970-491-8450, pierre@engr.colostate.edu

[2] Director Emeritus of the Fluid Mechanics and Wind Engineering Program, Department of Civil Engineering, Colorado State University, Fort Collins, CO 80523, 970-491-8574, meroney@engr.colostate.edu

small building, called the "claim shanty," was completed. According to tradition, it was built in eight days in order to prevent the site of the college from being moved to some other town. It is known that both Greeley and Boulder were circulating petitions to get the college away from Fort Collins.

Colorado was admitted to statehood in 1876, and the first State Legislature established the State Board of Agriculture in 1877, the governing body for the agricultural college. The Agricultural College of Colorado was to be governed by the Board consisting of eight members, at least four of whom had to be practicing farmers. It was about this time that a member of the legislature stated that money spent for an agricultural college would be money thrown away because Colorado would never be an agricultural state; "it was only fit for cow pasture and mining". Not until 1878 were funds for the College actually appropriated. Ground was broken for the "college building" later known as Old Main about June 20, 1878, and the formal laying of the cornerstone was on July 27 the same year at public ceremonies.

The Pioneers

Elwood Mead (1858-1936) was to play a pivotal role in hydraulics (Figure 1). The College originally employed him in 1882 as an instructor in mathematics, but he had studied civil engineering at Purdue and Iowa State. In 1883, Mead received approval to teach a two-term senior course on irrigation, one term to be "devoted to the pressure and flow of water, and methods of determining the same;" and the other "to the survey and construction of canals and reservoirs." That year Mead also became assistant state engineer doing practical fieldwork in irrigation. It is said that the State Board of Agriculture was interested in Mead because he would accept a position without the promise of a professorship the second year. In view of the shortage of funds, this could well be true. Mead evidently believed that the opportunity to serve in a needed capacity was more desirable than rank. He was immediately thrown almost literally and physically into the complicated problems of irrigation as they were developing in the Cache la Poudre Valley.

If any place ever needed irrigation, Colorado was it. Men fought and threatened to kill in order to obtain water for irrigation, and a way of measuring water so that each farmer could receive his fair share was urgently needed. Serving as assistant to the State Engineer E.S. Nettleton, Mead

Figure 1. Elwood Mead.

struggled with problems of measuring water, building ditches to direct water onto the land and storing water. Few laws existed at that time to control the use of water for irrigation, and before laws could be passed the concept of water as property had to be accepted. How was a man to acquire a right to water and to how much water? How was water to be measured? These were questions of great social and economic relevance.

In 1886, Mead became the first professor of irrigation engineering in the United States. He taught courses in "measurement and flow of water for irrigation." This was followed by work in hydraulics, canals, and dams. Surprisingly, one of their first class projects was not how to get water onto the land, but how to get it off. Working through head-high cattails, they surveyed and drained the swampy streams that crossed the campus.

It is said that President Ingersoll and the State Board of Agriculture created the professor position in an attempt to hold Mead on campus. At any rate, he left in 1888 to go to Wyoming where he wrote the irrigation code for that territory. As Wyoming's first State Engineer, Elwood Mead framed a revolutionary code of water law for arid and semiarid regions that was written into Wyoming's constitution and which became a model for irrigation laws adopted not only by four-fifths of the western states but also Canada, Australia, South Africa, and New Zealand. This new water law rejected the old English common-law principle of riparian rights as inappropriate for arid regions. Instead it declared all water, surface and underground, to be state property, thus giving them the same status as minerals and land and thereby ending legal conflicts between those who owned the land through which the water flowed and those who wished to use the water.

Mead was subsequently employed by the U.S. Department of Agriculture in Washington in 1899, the Australian Water Supply Commission, the University of California, and finally the U.S. Bureau of Reclamation in 1923. There is little doubt that the caliber of Dr. Mead's work was one reason for his appointment in 1924 as Chief of the Bureau of Reclamation. He served in this capacity for 12 years, until his death in 1936. F.D. Roosevelt said of Elwood Mead, "He was a builder with vision."

Mead's work as State Engineer provided him with an excellent knowledge of the terrain of northern Colorado and Wyoming. With a young assistant, he had surveyed for the first time much of Yellowstone. He knew "Buffalo Bill" Cody and in 1888 explored with him some of Wyoming's wild rivers. As commissioner of Reclamation he toured the West every year, and he was out west planning for a new project only a few weeks before his death. The lake above Hoover Dam now bears his name.

Engineering at the Agricultural College grew more rapidly than agriculture because of the high demand for engineers in irrigation. Also, engineering was a profession backed by the American Society of Civil Engineers, which was founded in 1852. Engineering had developed and tested principles it could rely and build upon, whereas agriculture did not have this foundation either nationally or in the state. A common attitude was "if my boy wants to be an engineer, I'll send him to college; but if he wants to be a farmer, I can teach him all he needs to know." About the turn of the century the emphasis of engineering over agriculture at the Agricultural College rose to the level of a controversy when an editorial writer pointed out that one student trained in irrigation engineering at the college was worth more to the state of Colorado than the cost of the college.

By 1903 the school had grown to an enrollment of 448. Included in the graduating class of 1904 in Civil and Irrigation Engineering (C&IE) was Ralph L. Parshall. Had the college kept a guest book during the early 1900's, many famous names would have been recorded. For example, in August 1904, a commissioner and a prince from Ceylon, and a representative from Egypt came to the college to learn about agriculture and irrigation. About the time the new C&IE building was first occupied, an expert sent by the Chilean government to study methods of irrigation and instruction in irrigation engineering in America, Egypt, Europe, and India reported that there was only one other institution in the world that had equipment for teaching and investigation of irrigation engineering equal to that of Colorado Agricultural College (CAC).

In 1910, the U.S. Department of Agriculture (USDA) stationed Victor M. Cone (1883-1970) at Fort Collins to take charge of U.S. Irrigation Investigations, an agency of the Bureau of Public Roads. This agency (forerunner of the Agricultural Research Service) in cooperation with the Colorado Agricultural Experiment Station was instrumental in building the new hydraulics laboratory of the C&IE department, and in 1912 Cone and Parshall helped in its design. The next year, Parshall was promoted to assistant professor, but he then resigned from the college to accept a position with the USDA, but remained in residence in the C&IE building. He was replaced at the college by Oliver P. Pennock (1879-1968), a rather reserved 1902 graduate who 40 years later became department head. In 1914, Carl H. Rohwer (1890-1958) of Nebraska and Cornell was transferred to Fort Collins by the USDA. Cone, Parshall, and Rohwer made the region, and vicariously the College, well recognized for irrigation research.

In 1920, Parshall and Rohwer made a search for an outdoor laboratory site, not too far from the city and with an ample supply of water. The waste gate on Jackson Ditch, leading from a branch of the Cache La Poudre River near Bellevue, northwest of Fort Collins, was found to meet their requirements, and a 7x14x75 foot concrete channel, tapering over another 50 feet to an outlet width of 25 feet, was connected to the gate. The latter permitted some adjustment to the flow, and a 15-foot weir was used for discharge measurement. It was in this channel that Parshall developed his adaptation of the Venturi flume for discharge measurement (Figure 2). Patented in 1922, it became widely known under his name and used around the world. Two of the advantages of the Parshall flume are: (1) it does not fill with silt, and (2) it is accurate, even on a relatively flat land where the velocity of water is low. Parshall is also known for a series of sand traps he designed to keep irrigation canals free of deposits.

Growth on Campus

In 1919, President Lory reported to the State Board of Agriculture that the college was out of debt, and between 1920 and 1930, eleven buildings were added to the campus. Local fraternity and sorority chapters began nationally affiliations around 1915. The first dormitory was remodeled for civil engineering and no new dormitories were built on campus until the construction of Rockwell Hall about 1939.

In 1919, Lory recommended that the College resume research into the possibility of diverting western slope snowmelt through the Continental Divide and onto the eastern plains. The drought and depression of the 1930's took their toll on the State, but Lory was instrumental in organizing support and lobbying Reclamation commissioner Mead for backing. With the help of Parshall and others, Lory then acquired a grant for the College from the Works Project Administration to conduct a feasibility study. By 1938, the construction of the $160 million Colorado-Big-Thompson River Project was underway.

In August 1930, the Bureau of Reclamation sent a dozen engineers, technicians, and shop people from Denver to Fort Collins to work in the laboratory, which had been designed by Cone and Parshall for the USDA. The Bureau program began with a study of proposed shaft spillways for Hoover Dam. As a result of these tests, the shaft was changed to side-channel structure. Thereafter, many other studies were undertaken, in particular for the Bureau's Grand Coulee and Imperial Dams and for the Tennessee Valley Authority's Wheeler and Norris Dams.

Figure 2. Ralph L. Parshall. (Rouse, 1976; reproduced by permission of IIHR)

Emory W. Lane (1891-1963), who had studied at Purdue and Cornell and who gained considerable experience both in the United States and in China, was administrative head of the Fort Collins operation. This involved duties at the laboratory during the Hoover spillway tests under Charles W. Thomas (1906-1978) and James W. Ball (1905 -...), both Coloradoans educated at Fort Collins. Lane later went back to Denver, turning the Fort Collins work over to Jacob E. Warnock (1903-1949), who held degrees from Purdue and Colorado. Upon Warnock's move to Denver, Ball was left in charge of the laboratory. By 1936, the laboratory had undergone a fourfold expansion, but for political and financial reasons the Bureau brought its work there to a close only two years later and withdrew to its Denver quarters in the New Customhouse. On his retirement from the Bureau of Reclamation in 1953, Emory Lane received a temporary appointment at Fort Collins, which he held until illness forced an end of his activities in 1957. By then he was well along in the formulation of a general philosophy of sediment transport.

The name of the Colorado Agricultural College was changed in 1935 to Colorado State College of Agriculture and Mechanic Arts (only to change again to

Colorado Agricultural and Mechanical College in 1944). Three years later in 1938, Pennock was replaced as Department Head and Dean by Nephi A. Christensen (1903 -...), a native of Utah who had just obtained a Caltech doctorate under Theodor von Karman and Robert T. Knapp. Christensen's first accomplishment was to gain accreditation (previously refused) of his three engineering departments by the Engineers Council for Professional Development. One of his former colleagues at Caltech was the Toledoan, Hunter Rouse (1906-1992), who had just become a professor at the State University of Iowa. He was invited by Christensen to give a 1940 summer class at Fort Collins in fluids mechanics. This attracted some two-dozen graduate students (among them J.C. Stevens, later president of the ASCE, and C.P. Better, sediment specialist of the Bureau of Reclamation), thus becoming the first of a continuing series of summer courses and conferences.

As the United States became involved in World War II, some college laboratories undertook war related research while other staff members moved to federal laboratories for similar work. Christensen played an important part in the development of rocketry at the Army's Aberdeen Proving Grounds, taking with him a number of the College staff, in particular Dwight Gunder (1905-1964), a professor of engineering mathematics. At the same time, Maurice L. Albertson (1918-...) with degrees from Iowa State College and the State University of Iowa, and who was to play a leading role in later developments at CSU, was called back from TVA to the Iowa Institute of Hydraulic Research for war related work under Rouse. This involved air tunnel tests on fog dispersal, turbulence, and jet diffusion, in the course of which he completed a doctoral dissertation on boundary-layer evaporation. Since he had long hoped to take part in the irrigation research at Colorado State, a position for him there was arranged in 1947.

At the time, Rouse opposed Albertson's move to CSU and warned him that if he'd move to CSU, nobody would ever hear of him again! Ironically, Rouse moved to Sun City, Arizona, and began teaching the summer course "Intermediate Fluid Mechanics" at CSU in 1976. Reminiscing Rouse's teaching, it quickly became clear that student involvement was insufficient for the study of fluid mechanics and that a full commitment was required of the students. Daryl Simons aptly clarified the difference between involvement and commitment, "if you have eggs and bacon for breakfast, the chicken was involved but the pig was committed." Hunter's summer teaching included a weekly classroom oral quiz where students answered various fluid mechanics questions in front of the class. After a student's answer, Rouse paused and said, "I have been teaching Fluid Mechanics for 50 years and this is the worst answer I ever heard." It is with pride that students were wearing "I survived Hunter Rouse" T-shirts in Fort Collins. Rouse retired in 1986.

Albertson contributed to the entrepreneurial spirit and served as catalyst for unprecedented research at the College of Engineering. At the same time that graduate engineering programs were implemented, the college embarked in the 1950's on international programs, thereby establishing one of the hallmark characteristics of the modern engineering program. The cold war battle for the loyalties of third world countries centered on the efforts on President Truman's Point IV program designed to aid the economic development of these third world regions. As part of this program, Colorado A&M became involved in a faculty exchange program with its sister school, the University of Peshawar in Pakistan. Four years later, the State Department approached Maury Albertson with a request for a feasibility study for a proposed graduate engineering school in Southeast Asia. With members of the Southeast Asia Treaty Organization (SEATO) seeking the region's economic and social development, the SEATO Graduate School of Engineering in Bangkok, Thailand, opened in the Fall on 1959 with Maury Albertson as campus coordinator. In 1967, the college turned into the Asian Institute of Technology. During the decade 1950-1960, most of the research activity in engineering at Fort Collins was within the Civil Engineering Department. Research reached a funding level of approximately $160,000 in 1955-56 and $277,000 in 1956-57. Contracts and grants provided most of the funding for research that exceeded $600,000 in 1961.

Growth as Colorado State University

College enrollment reached 1,000 in 1924, 2,000 in 1939, 6,000 in 1960 and 16,000 in 1969. In 1957 the institution's name changed to Colorado State University, and the academic divisions became colleges. T.H. Evans was Dean of the College of Engineering and D.F. Peterson was Head of the Civil Engineering Department. In 1958, M.E. Bender succeeded Peterson as Head of the Civil Engineering Department.

1957 was a memorable year, when the U.S. Geological Survey stationed Daryl B. Simons (1918-...) at Fort Collins to lead, with Everett V. Richardson (1924-...), the growing research program on river mechanics and sediment transport. Simons not only supervised the USGS program, but also completed work toward CSU's second engineering doctorate and taught courses in civil engineering. His primary contribution was the analysis of bedforms and resistance to flow in alluvial channels. He retired from CSU in 1983 and his expertise in sedimentation and river engineering has been sought throughout the world. In 1957, hydrologist Vujica Yevjevich (1913-...) came to CSU from the former Yugoslavia where he previously headed a research institute. The same year, A.R. Chamberlain, professor of civil engineering, was made chief of the Civil Engineering Section. This was a position created to administer the burgeoning research section of Civil Engineering.

When Chamberlain became vice president for administration in 1960, he left his position as chief of the research section, which vacant until 1963. At the same time, Bender left as head of the department to become Dean of the SEATO Graduate School of Engineering. J.W.M. (Bill) Fead became acting head of the Civil Engineering Department and D.B. Simons became acting head of the research section. Lionel Baldwin (1932-...), a chemical engineer with a specialty in fluid turbulence, came to campus in 1961 after experience with NACA-NASA. William W. Sayre (1927-...) of New York and Princeton, initially a graduate student, became a member of the USGS staff in 1962, moving to Iowa in 1968 after receiving the doctorate; his particular interest was the mechanics of diffusion. Hsieh-Wen Shen (1931...), a native of China who, after study at Michigan, had taken the doctorate in sediment transport under H.A. Einstein at Berkeley (and was to become an ASCE Freeman Scholar the following year), arrived at Fort Collins in 1964. Shen led numerous NSF projects in the field of sedimentation and river mechanics before returning to Berkeley in 1985. In 1965, Fead became Head of the Civil Engineering Department and Simons became the Associate Dean for Research for the College of

Engineering. This, of course, called for some reorganization of duties and responsibilities. Lionel Baldwin was appointed Dean of Engineering in 1965; and in 1969 Ray Chamberlain, for several years Vice President, assumed the post of President as the University was preparing to celebrate its centennial year.

In most respects the College of Engineering was well prepared to meet the demands of the boom years. Faculty expanded to more than 100 by the early 1970's while funded research skyrocketed from just under $500,000 to $5,000,000. With the state supplying as little as half of the funds needed to for instruction in some departments, it became incumbent upon faculty to prove the axiom that an engineer can do as much with one dollar as others with two. This reliance on soft money was not for the faint of heart. As Fead explained: "faculty had a lot of freedom in the way they operated and a lot of pressure to bring their own salary. But they were the type of people who wanted control of their own projects and who were willing to put up with the hassles of the entrepreneurship requirement to have that kind of working environment."

Growth at the Engineering Research Center

A new Engineering Research Center was built in 1962 at a cost of $1.8 million joining three wings together to provide 69,000 square feet of space for offices, conference rooms, small laboratories, electronics shops, printing, drafting, and photographic quarters, two lecture rooms, and a cafeteria. Directly south of the main wings and connected to them are two large laboratories, each roughly 120x280 feet in plan (Figure 3).

Permanent features in the Hydraulic Laboratory are a series of interconnected sumps 8 feet in depth and 5400 square feet in area; 14 pumps ranging in capacity from 250 gpm at 50-foot head to 23,000 gpm at 19-foot head; a 4x8x200 foot power-tilting flume with a discharge capacity of 100 cubic feet per second; a 20x100 foot river-basin flume for meander, erosion, and control-structure studies; a large local-scour flume; three other tilting flumes; and ample space for temporary models.

A 100-acre outdoor laboratory adjoins the building, and makes possible large-scale model and full-scale prototype studies. A 8x20x180 foot concrete flume with a

Figure 3. The Engineering Research Center from the top of Solider Canyon Dam.
(Rouse, 1976; reproduced by permission of IIHR)

recessed 10 feet deep section provides a facility for large-scale tests. A 3-foot-diameter variable-slope pipe 825 feet long is also available. A hydro-machinery facility is housed in a 70x192-foot prestressed-concrete building. The concrete floor slab is 3 feet thick to eliminate vibration during testing. Water for both the indoor and outdoor laboratories comes from the U.S. Bureau of Reclamation's Horsetooth Reservoir just uphill to the west of the Research Center.

A maximum head and discharge of 200 feet and 300 cubic feet per second are available, but modifications are currently underway that will increase the maximum available discharge to 500 cubic feet per second. In addition, large stationary and movable pumps are used to recirculate water from sumps and to increase the operating head.

With such facilities and staff, not to mention the considerable support of various agencies, graduate enrollment steadily rose. In of September 1979 the staff of the combined hydraulics, fluid mechanics, and hydrology sections of the Civil Engineering Department had grown to a total membership of 125. Of these, 33 were of faculty rank, i.e., assistant professor or above, and 92 were graduate assistants. The hydraulics section was the largest, with 69 members, and fluid mechanics next with 36. About 55% of all graduate students in the Department were employed part-time, with an annual turnover of some 40-50%.

Fluid Mechanics and Wind Engineering

The beginning of the Fluid Dynamics and Diffusion Laboratory (FDDL) [currently called the Wind Engineering and Fluids Laboratory, WEFL] was in 1949 when the first wind tunnel was designed and largely build by J.E. Cermak, who was at that time an instructor in the civil engineering department (Figure 4). The wind tunnel was located in the old Industrial Research Building, now called the General Services Building. Cermak and Albertson, at that time an associate professor in the civil engineering department, worked on the first research contract for the FDDL, which was a grant from the Office of Naval Research, 1949-54, for a study on evaporation.

In 1955 the Air Force granted funds for a Meteorological Wind Tunnel, and this became operational alongside the first tunnel in the new Engineering Research Center in 1963. The Meteorological Wind Tunnel became the first wind engineering facility in the world to include a long test section in which a simulated atmospheric boundary layer grew naturally, and in which heated/cooled boundaries and thermal controls permitted simulation of day time convection and night time inversion conditions.

The Wind Engineering and Fluid Mechanics Program under the direction of Dr. Jack E. Cermak grew substantially during the late 1960s, 70s and 80s until the program included some 12 faculty and research scientists and their graduate students. A third large facility, the Environmental Wind Tunnel was added with a test section 57 feet long and a 12x8 foot cross section to investigate wind effects over large terrain areas. During this time, Wind Engineering became a recognized engineering discipline, and in 1989 the National Society of Professional Engineers declared the CSU Wind Engineering and Fluids Laboratory one of the five major engineering accomplishments of the decade.

Several facilities for specialized research augment the larger test facilities, including a Separated-Flow Facility that has a working section 2-foot wide with a flexible floor that can be adjusted over a length of 10 feet to provide a wide range of pressure variation in the flow direction. Aerosol dispersion can be studied in an Aerosol Test Facility with a 2x2 foot test section 15 feet long capable of producing air speeds up to 116 miles per hour and equipped with a remote-sensing laser-powered particle spectrometer.

Figure 4. Jack Cermack. (Rouse, 1976; reproduced by permission of IIHR)

Studies of large-scale turbulence are made in a Gust Tunnel with a 3x3-foot test section equipped with two banks of airfoils whose pitch may be varied randomly by an electromechanical servo-system.

During the early 1960s models of the New York World Trade Center were examined in the Meteorological Wind Tunnel. Based on these measurements the original locations of the twin towers were reoriented, and a passive damping system was designed and installed in the towers to mitigate tower sway. Other early wind flow and loading studies included evaluations of wind effects at the Candlestick Ball Park, San Francisco, and the Oakland-Alameda County Coliseum. After the famous cladding failures that occurred in 1972 during the construction of the John Hancock building in Boston, Massachusetts, architect and design firms requested preventive model studies from CSU engineers, which resulted in many hundreds of model skyscraper studies in the CSU facilities during the 1970s and 1980s. In the late 1980s a model of the world's tallest building, the Chicago Sears Tower, was evaluated for wind effects to mitigate existing problems with glass and cladding damage.

In 1987, Robert N. Meroney, graduate of the University of California, Berkeley, led the program and emphasized research in areas of atmospheric transport of dense and buoyant gases, wind energy conversion, wind turbine siting, and computational fluid mechanics. He joined with Kishor Mehta of Texas Tech University and the two schools obtained the first multi-year US National Science Foundation Cooperative Program Grant. The grant focused on Wind Engineering and lasted from 1987 through 2001. This research program was directed to evaluate the hazards of extreme winds acting on low-rise buildings such as homes, churches, schools and shopping centers. The work resulted in a major revamping of the US National Wind Loading Design Codes, new specifications for extreme wind hazards for the continental United States, the creation of a major data base of field and laboratory wind load data, the incorporation of wind hazards into building insurability criteria, and a new understanding of the role of corner vortices in the wind loading of civil engineering structures. Other research associated with this grant included simulation of infiltration and exfiltration across building envelopes, and wind blown debris. Other contributors to the program include Jon A. Peterka, Willy Z Sadeh, and Virgil A. Sandborn.

Recent Years

In the 1980's, E.V. Richardson was leading the development of water resources at the International level and particularly in Egypt where CSU trained a generation of scientists and engineers with expertise in hydraulics and water resources. This successful program brought in about $60,000,000 to CSU from 1977-89. After his retirement in 1989, Richardson remained very active in the field of bridge hydraulics and led numerous projects on pier scour and stream stability for the Federal Highway Administration. Daniel K. Sunada took over the Egypt Water Research Center program that brought an additional $20,000,000 to CSU from 1989-93.

In the past two decades, the Hydraulics Laboratory has seen the construction and analysis of numerous large-scale physical models including several dams, spillways, energy dissipators, river models, analyses of resistance to flow, sediment transport, riprap and block stability, filter design, river morphology, local scour, bridge hydraulics, environmental hydraulics, surface runoff and sheet erosion, infiltration and contaminant transport, mudflow and debris flows, dam break, reservoir sedimentation, etc. Primary contributors to the Hydraulics Laboratory at the Engineering Research Center in the past several decades include Everett V. Richardson, Susumu Karaki, James F. Ruff, Stanley A. Schumm, Hsieh Wen Shen, Steven R. Abt, Pierre Y. Julien, Carl F. Nordin, Albert Molinas, Chester W. Watson, and many others.

During the same period, the Wind Engineering and Fluids Laboratory saw studies examining hazards associated with power plant pollution, siting of nuclear power stations, dispersion of dense gases due to chemical or liquified gas spills, flow in agricultural crops, bridge aerodynamics, pedestrian comfort around buildings, wind effects on space rocket launch facilities, snow drifting, odor pollution, wind turbine design and wind turbine facility siting, wind loads on solar collectors, etc. Currently the Wind Engineering and Fluids Laboratory is under the direction of Bogusz Bienkiewicz where he supervised the study of tornado dynamics in two new simulation facilities and evaluated wind effects on roofs and roof top pavers. David E. Neff, who continues the tradition of basic and applied research focused on wind engineering problems, manages the lab.

The past decade or so has seen the renovation of the Engineering Building at a cost of approximately $18,000,000. Recently created Civil Engineering Department faculty position include environmental hydraulics, environmental engineering, an Endowed Chair in urban hydraulics, hydrology, hydraulics and water resources. There are currently about 35 FTE faculty positions in Civil Engineering and about 25 faculty members work in water and fluid mechanics.

Summary and Conclusions

With roots dating to 1870 and a foundation in 1957, the great success and rapid growth that CSU has experienced in a short period of time can be attributed to several factors. These certainly include: (1) commitment of faculty members who could build with vision; (2) endless energy and productivity of faculty, students and staff; (3) collaborative research with numerous governmental agencies, peer institutions, consulting firms, etc.; and (4) active involvement and visibility at the national and international level. In conclusion, it can be said of CSU that its past was bright and the future...even better!

Acknowledgments

The authors, both non-historians, acknowledge delving into existing historical reports of the institution. The authors of the following references deserve the credit for the historical research, while we assume the blame for any important omission in the preparation of this manuscript.

References

Albertson, M.L. and C.N. Papadakis (1986), *Megatrends in Hydraulic Engineering - A commemorative Volume Honoring Hunter Rouse*, Colorado State University, Department of Civil Engineering, 468p.

Anderson, (1970), *History of the Department of Civil Engineering*, Unpublished report at CSU.

Cermak, J.E. and M.L. Albertson, (1958), "Use of wind tunnels in the study of atmospheric phenomenon," *Air Pollution Control Association Annual Meeting*, Paper No. 58-32, Philadelphia, PA, May.

Hilfinger, A. (1989), *One hundred years of engineering at Colorado State University: Fulfilling the Land-grant Mission*, College of Engineering, Colorado State University, 75p.

Rouse, H., (1976), *Hydraulics in the United States 1776-1976*, Iowa Institute of Hydraulic Research, The University of Iowa, Iowa City, 238 p.

Rouse, H. (1980), *Hydraulics, Fluid Mechanics and Hydrology at Colorado State University, Engineering Research Center*, Colorado State University, Fort Collins, Colorado, 87p.

"The Practical Scholar - The Life of Elwood Mead", Unpublished Report at CSU.

The St. Anthony Falls Laboratory In History

Ed Silberman, Roger Arndt, Gary Parker, Efi Foufoula-Georgiou, and Chris Paola[1]

Abstract

The laboratory, known until recently as the St. Anthony Falls Hydraulic Laboratory, was designed and built under the direction of a dedicated individual, Lorenz G. Straub. Straub had been a Freeman Fellow and observed several laboratories in Germany during the year of his fellowship. He came to the University in 1930 and promptly set to work to establish his own laboratory. His vision came to fruition through a WPA grant to the University of Minnesota and construction started in 1936. Straub came to be known as the "River Doctor" for his many studies at SAFL on several aspects of river engineering.

The Laboratory building lies on the Falls of St. Anthony in Minneapolis, Minnesota where there is a drop of about 15 meters. Up to 9 m^3/s may be drawn through the building and distributed to the many flumes for experimental research. Laboratory personnel have pursued studies in many areas of river engineering, hydrology and experimental and theoretical fluid mechanics. The legacy left by Straub is more than just a laboratory building and the equipment it contains. His vision of a university laboratory as a leader in the advancement of pioneering methods in water resources engineering as well as being an educational tool lives on. SAFL continues to explore cutting edge research on environmental and geophysical fluid dynamics and apply its knowledge to a variety of water-related engineering problems. This is a brief account of SAFL's history, present and future written by five of its Directors dating back to 1963.

Introduction

The year 2003 marks the 65th anniversary of the St. Anthony Falls Laboratory. This paper is intended as a summary of the Laboratory's accomplishments from its beginnings, as seen from a more distant viewpoint than in the references cited, and to recount its current course and possible future roles. A detailed early history of the Laboratory and its personnel may be found in the book by Mary Marsh, *The St. Anthony Falls Hydraulic Laboratory, the First Fifty Years* (Marsh, 1987).

The Laboratory was designed and built under the direction of a dedicated individual, Lorenz G. Straub. The Laboratory story cannot be told without knowing something about Straub. So, we begin with Straub and follow with the Laboratory

---

[1] St. Anthony Falls Laboratory, University of Minnesota, Minneapolis, MN 55414.

history up to the present. We are proud that the preparation of this paper has been a joint effort between the current Director (Foufoula), the newly appointed director (Paola) and three former Directors (Silberman, Arndt and Parker), dating back to 1963.

Straub

Lorenz G. Straub (Figure 1), a native of Kansas City, Missouri, obtained degrees, including the Ph.D. in 1927, in Civil Engineering (majoring in structural engineering) from the University of Illinois. In the spring of 1927, the flood of record to that time occurred on the Mississippi River; as a consequence, there was both public and engineering interest in designing river control works. At the same time, John R. Freeman, a prominent and successful hydraulic engineer, endowed a fund to enable American engineers to travel to hydraulic laboratories in Europe where research in river control (and other related topics) was more advanced than in the United States. These traveling fellowships were awarded through three societies - The American Society of Civil Engineers (ASCE), The Boston Society of Civil Engineers, and The American Society of Mechanical Engineers. Rouse (1976) provides more details, but ASCE awarded the first of these to Straub. He spent two years at several German laboratories and traveling in Europe.

Figure 1. Lorenz G. Straub

On his return to the United States in 1929, he was employed by the U.S. Army Corps of Engineers at its Kansas City district office to work on problems in hydraulics and sedimentation of the Missouri River. In early 1930, he was sought out by the University of Minnesota. In the fall of that year he accepted an appointment as Associate Professor of Hydraulic Engineering in the Department of Mathematics and Mechanics!

Marsh, who became Straub's personal secretary in 1955 and retired as Laboratory Administrator in 1981, cites evidence (Marsh, 1987) that Straub's position was created not to just fill a vacancy, but with the intent that the University would build a hydraulic laboratory and form a Department to teach and do research in hydraulic engineering and related areas of fluid mechanics. Straub, of course, was well qualified for this appointment by his sojourn in Europe and work for the Corps of Engineers. During negotiations leading to his employment, he was shown several reports prepared over a period of many years promoting the use of the falls at St. Anthony for building a laboratory. Marsh (1987) also documents that Straub had already started a design for a laboratory to be built at Kansas City for the Corps of Engineers.

The possibility of building a laboratory was derailed by the depression of that era. In the mean time, Straub taught and did research in an existing small laboratory in the University's Experimental Engineering building. In addition to the undergraduate "Hydraulics" class required of all engineering students, he taught graduate classes titled "Open Channel Flow", "Mechanics of Similitude", and "Mechanics of Sediment Transport". Straub's research was largely on transport in open channels and in porous media and in stability of sand dams. He also found time for some consulting. He encouraged several of his graduate students to undertake experimental theses in the then more modern areas of fluid mechanics like viscous flow in channels. In fact, all his graduate students were expected to take the few courses that were available in the mathematics curriculum dealing with hydrodynamic theory and practice using the books by Lamb (1932) and Prandtl and Tietjens (1934).

Finally, in 1935, the federal government created the Works Progress Administration (WPA) to reemploy the millions of idle workers in the country. The Dean of the College of Engineering immediately commissioned Straub to investigate the possibility of using that program to build a laboratory at St. Anthony Falls. Through many tribulations, the St. Anthony Falls Hydraulic Laboratory was built and dedicated on November 17, 1938, but Straub was not officially named its Director until the spring of 1942.

Straub designed the laboratory and all the equipment that went into it at the time; he fought many battles with WPA and with the University over the amount of money he required; he faced labor strife; even his health suffered (he was a diabetic). Following dedication of the Laboratory, Straub continued his teaching, research, and consulting; he also struggled, with some success, to bring in sponsored experimental research to support the Laboratory. He encouraged one of his doctoral students to undertake experiments on surface air entrainment in high velocity flume flow. This continued as an important research area well into the 1950's and resulted in publication of a prize-winning paper (Straub and Anderson, 1958). Another student of that period, John S. McNown who obtained his Ph.D. in 1942, later wrote a very perceptive article about Straub (McNown, 1992).

However, before a fully meaningful program could be established at the Laboratory, Straub and most of his assistants were called to participate in civilian activities related to World War II. Straub went to New York City to serve on the National Defense Research Committee. He did not return until January of 1945, although he made occasional visits.

Straub returned to the University after prolonged negotiations with the administration as to his and the Laboratory's status. He had offers from other universities, and he used these as leverage. He wanted an almost autonomous Laboratory and European type "chair" within what had become the "Institute of Technology". He finally agreed to become head of a newly created Department of Civil and Hydraulic Engineering and to the title, Professor of Civil Engineering and Director of the St. Anthony Falls Hydraulic Laboratory. He also brought to the Laboratory three of his former graduate students (Alvin G. Anderson, John F. Ripken, and Edward Silberman) to lead the research program and to do much of the teaching, all under his close supervision. No research could be undertaken, no courses could be taught, and no outside technical activities could be undertaken, without his approval.

Straub carried on an intensive consulting practice after his return, in addition to doing some teaching and administering the research program at the Laboratory and running the Civil and Hydraulic Engineering Department. He reimbursed the University for 25 percent of his salary to permit the consulting work, but that work brought in many hydraulic model studies that supported the general Laboratory budget handsomely. He became recognized internationally for his ability to diagnose and recommend solutions to hydraulic engineering problems and was dubbed the "River Doctor" by a national magazine.

Finally, his diabetes and intense work regime caught up with him and a graduate student found him dead at his desk early Monday morning, October 28, 1963. Two memorials exist in his name at the Laboratory. First, is the Lorenz G. Straub Memorial Library. Second, an endowment fund was established by contributions from his numerous colleagues and industrial contacts. This has been devoted to supporting the Lorenz G. Straub award. The award is made annually to the author of an outstanding Ph.D. thesis in hydraulics or a closely related area.

Straub's Laboratory (1938-1963)

St. Anthony Falls on the Mississippi River (Figure 2) was first suggested as a laboratory site in 1908. The head of Civil Engineering at the time wrote to the Dean of the College of Engineering that the city of Minneapolis water supply plant located on Hennepin Island at the Falls had been abandoned a few years earlier (following a typhoid epidemic) and that this would be an excellent place to develop a hydraulic laboratory (Figure 3) for the Department. It was noted that there was 50 feet (15 meters) of hydraulic head and that water rights would be acquired with the site, which was located just two miles from the University's Engineering building. Nothing was produced by this letter or by several studies and reports during the following years until the hiring of Straub in 1930. Most of the early reports emphasized that a laboratory would be used to study pumps and turbines, and that support could be expected from manufacturers of those machines as well as from a potential hydropower industry in Minnesota. Also noted was the potential for developing and calibrating instruments and making river models.

The geology at the site consisted of earth and boulder overburden, a 20 ft layer of fairly sound Platteville limestone that was exposed at the river bluff, and under that, the friable St. Peter sandstone.

Figure 2. St. Anthony Falls

The main experimental floor of the building was dug into the limestone so that it would lie about 20 feet below the river level on the upstream side of the falls. The Laboratory is well described by Straub in the following paragraphs copied from Marsh (1987).

Figure 3. Artist's Sketch of the Laboratory, 1937

"The laboratory may be divided into essentially five units, the main experimental laboratory, the hydraulic machinery laboratory, the turbine testing laboratory, the large-scale volumetric measuring tanks, and the administration and lecture rooms."

"The main experimental laboratory is approximately 300 feet long and 45 feet wide. It is two stories high and contains three large channels extending the entire length of the structure. One is an overhead flume 8 ft wide and 9 feet deep, connected directly to the headwater above the falls, and is provided with numerous off-takes to supply water for the various experimental projects. The others are low-level channels below the level of the main floor. Of these, one is a wasteway and the other an experimental flume arranged for a wide variety of experiments. It is 9 feet wide and 6 feet deep, and is supplied directly from the upper pool of the river through a pressure tunnel. Enough head is available to put water through the flume at the rate of about 35 cubic feet per second for shallow depths. A towing car will make it possible to pull current meters, model ships, and the like through the flume with the water either at rest or in motion."

"The hydraulic machinery laboratory also has a clear height of two stories and is 34 feet wide and 125 feet long. It will be provided with an overhead crane. At one end of this laboratory there is a penstock shaft about 20 feet square and 30 feet deep below the machinery testing floor; the shaft provides a means of bringing the water from the overhead channel to the turbine testing laboratory, the floor of which is about 46 feet below the headwater pool."

"The turbine testing laboratory adjoins the hydraulic machinery laboratory, but at a lower level. It is of irregular shape in plan, two sides being formed by the limestone ledge of St. Anthony Falls. The turbine laboratory is approximately 60 feet long and 75 feet wide. A tailrace channel traverses the length of the laboratory beginning in the penstock shaft and extending to the tailwater pool below St. Anthony Falls."

"The volumetric measuring basins are constructed with their bottom just above the tailwater pool. They are so located that the flow from all parts of the laboratory except the turbine-testing laboratory can be measured. A central control house is arranged to operate large cylindrical valves in a diverter system for the tanks, also in the tanks themselves. Recording and indicator gages will be located in the control house. The valves are laid out to operate pneumatically. The measuring system is designed to handle a continuous flow up to 300 cubic feet per second. It is intended to use this discharge measuring arrangement for checking measurements on large scale experiments and for calibrating water measuring devices to be used in connection with the turbine testing laboratory."

"Administration and lecture rooms are provided in a superstructure above the hydraulic machinery laboratory. A unique feature here is a demonstration lecture room so arranged that large quantities of water could be readily handled in various types of demonstrational experiments. Below the lecture platform is the main overhead supply flume for the laboratories while above the lecture platform is a head-control room containing a constant level reservoir. At one side of the lecture platform is a stairwell and pipe shaft providing access to the laboratory below and the control room above; at the other side, an apparatus room is arranged for housing the various pieces of demonstrational equipment."

"Access to the laboratory is by means of a roadway which bridges over the head race to an adjoining power plant and also over the roof of the main experimental laboratory down a ramp to the main level about 20 feet above the headwater pool."

It might be mentioned that the penstock, described above, which brings water to the turbine-testing laboratory was the original shaft for the Minneapolis water supply. Also, all of the flow structures described above were modeled in the Experimental Engineering laboratory space assigned to Straub.

A few of the things Straub planned did not materialize, but much was added to the Laboratory under his direction during his tenure. Both Ripken and C. Edward Bowers, also a retired professor, were his principal assistants in the design of both the building additions and major items of equipment.

As to the building structure, several internal floors were added, sometimes as mezzanines, reducing two-story spaces to single stories; an elevator was installed connecting the original upper floors and the turbine room using the stairwell at the downstream end of the building and excavating below that to the lowest level; and, just before his death, a new floor was completed above the roof of the main experimental laboratory and adjacent to the administrative floor. One end of this new structure was used to house the Lorenz G. Straub Memorial Library.

Straub learned in early 1938, before the Laboratory was completed, that the St. Paul District Office of the Corps of Engineers intended to perform a model study in connection with extending navigation on the Mississippi River so as to create an upper harbor in Minneapolis above the Falls. The District Office maintained a sub-office at the Iowa Institute of Hydraulic Research with personnel who regularly performed river model studies. Straub fought a battle, which extended to the halls of Congress and eventually forced the Iowa personnel to come to St. Anthony Falls to build the model! Straub then faced another battle to convince the University to provide money to build a floor where the model (Figure 4) could be placed. He managed to do this, too. The new floor covered the entire main experimental hall, dividing its height in two, and when the model was completed, it covered most of that floor. This space was named the River Model Floor and has served to this date as the location for a great number of other models.

Figure 4. Original Mississippi River Model

At this point, something needs to be said about financing. In his original agreement with the University to return as Head of Civil Engineering and Director of the Laboratory in 1945, Straub extracted several commitments from the University administration (through the Dean of the Institute of Technology). A technician, maintenance man, and part time secretary serving the Laboratory were to be carried on the Civil Engineering Department payroll, and he was authorized to establish an independent revolving fund at the Laboratory, funded using receipts from indirect costs on research projects and from reversions on the salaries of academic staff members who worked on research during the school year and thereby had a portion of their salaries paid by the projects. At least twice during his tenure, Straub had to battle the administration to maintain this agreement. Thus, the Laboratory was nominally "self-supporting" so that building additions and new major equipment were readily financed.

Several major items of equipment were added under Straub's direction, associated with Laboratory research programs. A large tilting flume replaced the original one used in the air entrainment research, already mentioned (Straub and Anderson, 1958) (and was later dismantled). Research on several aspects of cavitation resulted in construction of a six-inch water tunnel (Straub et al., 1955) (which was originally designed as a model for a sixty-inch tunnel for the David Taylor Model Basin of the U.S. Department of the Navy), a ten-inch free jet gravity-flow water tunnel (Silberman and Ripken, 1959), later supplemented with a two-dimensional test section, and the development of a towing carriage for the nine-foot wide main channel in the experimental laboratory (Straub and Bowers, 1956). Both a mechanical and a pneumatic wave maker and beaches of various types were designed for this channel to support research in the wave making process itself, as well as to test structures (both rigid and erodible) in waves. The channel and towing carriage could then be used to test ship models, hydrofoils, and other watercraft in waves. Considerable effort was given to instrument development for use in these facilities; John M. Killen, a research fellow now retired, was trusted to lead this effort.

Straub built river models and river structure models (dam sections of earth, rock, and concrete, similar spillway sections, gates of many types, fish ladders, coffer dams, open and closed water conduits) wherever space could be found in the building. This is the main reason mezzanines were constructed in all of the original two-story parts of the building. Numerous other areas of research, both basic and applied, were addressed: air-water mixture flow, including acoustics, hydrofoil development, non-Newtonian fluid flow, and boundary layers. The Laboratory list of publications (through 1963) may be examined to obtain a more complete picture of the program during that time.

The transfer of the Mississippi River model to St. Anthony Falls eventually resulted in the transfer of the entire Corps sub-office to the Laboratory where it remained long after Straub's death. Another federal research group was also established at the Laboratory about this time and, again, required intense negotiations by Straub. This was the research arm of the Soil Conservation Service, which later became the Agricultural Research Service (ARS) whose long time head at the Laboratory was Fred W. Blaisdell. This office, too, remained until well after Straub's death. Later, the Federal Interagency Sedimentation Project was also established at the Laboratory with personnel from the U. S. Geological Survey as well as from the sub-office of the Corps of Engineers.

Straub established a series of publications for the Laboratory (SAFL, 1981). There were Project Reports, limited distribution publications giving complete details of each project reported upon and intended for the sponsor and as a permanent record; there were Technical Papers in two categories, A and B, the former being reprints from peer-reviewed publications and papers presented at conferences and elsewhere and the latter which were published only by the Laboratory after review within the Laboratory, describing research not published or presented elsewhere. There were Circulars dealing with general information about the Laboratory, and there were motion pictures. Every publication bore Straub's name, at least to state that he was Director of the Laboratory. Straub did not encourage publication outside the Laboratory. These categories are still in use, save for the Technical Paper series B.

The Laboratory After Straub (1963-1999)

On the evening of Straub's death, Edward Silberman was called by the Dean of the Institute of Technology and asked to take the position of acting Director of the Laboratory. Subsequently, he became Director without further formal action. He resigned that position on June 30, 1974 and was replaced by Alvin G. Anderson. Anderson died in office after exactly one year. John F. Ripken then became acting Director while a search committee sought a new Director. Roger E. A. Arndt was selected for the position and took office in January 1977. Arndt was the first Director who was not a student of Straub; but he had a connection to the Laboratory - he was the third recipient of the Lorenz G. Straub award, in 1968. Following Arndt, Gary Parker assumed the Directorship in 1995. Parker had been a student of Anderson and was thus a second-generation student of Straub. Finally, in 1999, a new Director, Efi Foufoula-Georgiou, who had no previous connections with the Laboratory, was selected to replace Parker and is the current Director. A newly appointed director, Chris Paola (currently, the co-Director), will assume responsibility in September 2003. Paola, who is a faculty in the Department of Geology and Geophysics, will be the first director who is not a faculty of the Department of Civil Engineering. This is a vivid demonstration of the natural broadening of the Laboratory's breadth over the past decade to areas of geophysical fluid dynamics and earth surface processes. This evolution will be discussed in a later section.

During Silberman's tenure as Director, the Laboratory program was aimed at intensifying the already robust naval hydrodynamic research (cavitation, hydrofoil development, underwater noise, and the like); obtaining more National Science Foundation (NSF) support for basic research in the areas already supported by the Navy and in new research in stratified flows, turbulence, and hydrology; and maintaining the river and hydraulic structure modeling work at which Straub had been so successful. An unsuccessful attempt was made to begin a program in wind engineering (a wind engineering facility was later built under Arndt's directorship).

In a brief time, the modeling work, which brought considerably more indirect cost reimbursement to the Laboratory than was obtainable from government agencies, fell off considerably. After a few years, support by U.S. Navy research agencies was also reduced by acts of Congress. Furthermore, Silberman was unable to maintain the agreements Straub was able to extract. The civil service positions in the Civil Engineering budget were dropped right after Straub's death. Faculty salary reversions ceased within two years. By 1969, when a number of staff people had to be discharged, the Dean of the Institute of Technology attempted to arrange a sale of the Laboratory to a private research corporation. Fortunately, just at this time, the President of the University returned from a trip to several South American Universities. Within days of his return, he called Silberman to arrange a visit to the Laboratory and stated that the reason for his visit was to observe for himself what many of his hosts on the trip had told him about that "wonderful installation at St. Anthony Falls." The proposed sale ended there. Subsequent adjustments had to be made so that the University obtained some of the indirect costs, but financial problems were alleviated.

Added to the Laboratory during Silberman's tenure was the Buoyant-Body Test Facility, a prominent vertical 60-inch pipe structure rising from the turbine laboratory floor through its roof and then outside past the machinery laboratory and administrative floor to above the building roof (Killen, 1974). Another addition was a large, high-speed water tunnel (SAFL, Circular 5) installed on a mezzanine floor in the turbine laboratory. And, this being the start of the mini-computer age, a stand-alone computer was purchased so that punch cards would no longer have to be carried to the campus for less demanding tasks. But there were no building modifications.

Silberman was more successful in maintaining the academic structure Straub had created. During the 1950's, what became the Department of Aerospace Engineering and Mechanics established its own fluid mechanics courses. Within weeks of Straub's death, the department head requested that the Dean immediately transfer all fluid mechanics courses to his Department. Mediation resulted in preserving the status quo and led, eventually, to complete cooperation between faculty members of the two groups as well as with those in Mechanical Engineering, Chemical Engineering, and Mathematics.

In addition to administering the Laboratory and teaching courses in fluid mechanics, Silberman became interested in Water Resources Management and later introduced a course with that title. When Congress passed the Water Resources

Research Act of 1964, creating Water Resources Research Centers in each State, Silberman attempted to have the University build a new floor over the administrative level of the Laboratory so that the Minnesota Water Resources Center could be housed there. He was not successful in this. But the activity created by the Act resulted in the formation of the American Water Resources Association, which Silberman joined as a charter member. In 1969, he became President of the Association and in 1974, at the end of his tenure as Director, the Association moved its headquarters to the Laboratory (to remain until 1982).

Silberman resigned from the Directorship following an amicable disagreement with the head of the Department of Civil Engineering over the assignment of faculty to the Laboratory. It had been agreed in advance that Dr. Alvin G. Anderson would become Director on July 1, 1974. No search was conducted among possible candidates because of limited financial resources to fund the position. Anderson had completed his Ph.D. dissertation in 1950 under Silberman's guidance.

One of the obvious changes Anderson set in motion almost immediately was to stress research in more traditional Civil Engineering areas, especially sediment transport and erosion, at the cost of reduced support for naval hydrodynamics. This was apparent in support for travel to organizational conferences. In fact, Anderson died in trying to reestablish a prominent place for the Laboratory in the International Association for Hydraulic Research, an organization which Straub had help to found and of which Straub had been President. Anderson was en route to the biennial congress of that organization at his death on July 1, 1975.

Anderson had little time to make a large mark on the Laboratory building or equipment, but he is best remembered for the graduate students he counseled. After his death, family, colleagues, and former students established a memorial fund in his name and this has been used to fund the Alvin G. Anderson Award. The Award is made to a graduate student at the Laboratory working in Anderson's major field of interest, sediment transport, or a related area. The first award was made in 1976 to S. Dhamotharan, who is now a Senior Vice President at URS Corporation.

On Anderson's death, Silberman declined an offer to be an interim Director while a search was conducted for a new Director. Professor John F. Ripken took on this task and guided the Laboratory as much as was possible on the course Straub had set and Silberman had tried to follow. Ripken was personally interested in several aspects of cavitation research and published regularly on this subject. He left no special mark during his Directorship, but his name was associated with the Laboratory in important respects from the original building design through the date of his retirement in 1979. He now lives in retirement in Minneapolis.

Dr. Roger E. A. Arndt was named as the new Director near the end of 1976. Arndt came to the Laboratory on a part time basis in January 1977 and assumed full time duties that summer. He came to the position with a different background and a different perspective. Although educated as a civil engineer with graduate work at MIT under Daily (SM) and Ippen (PhD) his industrial background and prior academic post at the Pennsylvania State University were in aerospace engineering.

Arndt's first objective was to revitalize the fundamental research program and to increase graduate student participation in the operation of the laboratory. He felt that the time was ripe to base both the fundamental *and* applied work on a firm foundation of fluid mechanics. As shown in an early presentation, his view of the laboratory was one of an integrated program of education, fundamental research and service to the profession (Figure 5). It was expected that students would benefit from involvement in both basic research where grounding in fundamental principles would be received and, where feasible, participating in applied research where real world problems are solved within time and budget constraints. The basic program would benefit from the identification of relevant problems for future research during the course of an applied study. The applied work would in turn benefit from the use of the latest experimental and computational methodology developed in the basic program. The vision was that the integration of the three objectives would benefit greatly from this synergism. Education, research, and applications are all integral parts of a major research university.

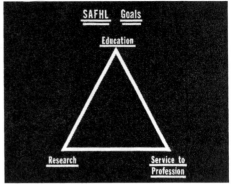

Figure 5. SAFL Goals

To commemorate the 40th anniversary of the laboratory, Arndt and Marsh organized the Symposium on Fluid Mechanics in Water Resources Engineering, which was held in April 1979 (Arndt and Marsh, 1981). The Symposium addressed fluids mechanics research in water resources engineering, both as to the state-of-the-art and the future directions of fluid mechanics. This symposium served as a road map for further development of the laboratory. Following the symposium, the Laboratory underwent substantial change. The Laboratory had built its reputation on the use of hydraulic models for solution of a variety of hydraulic and river engineering problems, the development of specialized instrumentation, and naval hydrodynamics research. This activity was expanded to include topics such as computer simulation of water quality dynamics, thermal pollution, aeration, fluid transients, slurry transport, acoustic radiation from bubbly flows, cavitation, hydropower, ice formation, surface and groundwater hydrology, and the design of major research facilities.

The new hydro-turbine research facility and the new boundary layer wind tunnel, both of which are unique installations, considerably enhanced the Laboratory's capability in experimental research. Computer simulation of everything from physico-chemical processes in rivers and lakes to flow through the turning vanes of a $150,000,000 water tunnel became almost routine, mostly under the direction of Charles Song, now an Emeritus Professor. The Laboratory's new linkage with the University's Supercomputer Institute provided the opportunity to meet the challenges during this period of rapid expansion.

Much of the expansion was based on several new opportunities that presented themselves during this period. The energy crisis of the late 1970's fostered an interest in small hydropower, both on the Federal and State level. The Legislative

Commission for Minnesota Resources (LCMR) began to recognize the University and SAFHL as a significant resource for appropriate research in water resources and energy. The US Navy saw the need for considerably expanded research in naval hydrodynamics, hydroacoustics and cavitation while the National Science Foundation became interested in developing a program of fundamental research to complement a growing activity of applied wind engineering being carried out at other laboratories.

Probably the most significant opportunity was the appointment of several new faculty, following the appointment of Heinz Stefan in 1967. The new faculty were Cesar Farell in 1978, Gary Parker and John Gulliver in 1980, Peter Kitanidis in 1984, and Efi Foufoula-Georgiou in 1989. Farell brought expertise in wind engineering and fluid mechanics. Parker was key to revitalizing research in river mechanics and fluvial hydraulics and in developing a coordinated effort in both fundamental and applied activities in this area. Gulliver brought a new perspective to environmental research and was key to adding *engineering solutions* as part of the water quality research program. He also played an active role in all phases of hydropower research, and was instrumental in developing stronger ties with the environmental faculty with the Civil Engineering Department. Kitanidis (who was hired to replace Edward Bowers in the area of hydrology) remained at Minnesota only briefly and left to assume a faculty position at Stanford. Efi Foufoula-Georgiou, who developed an active research program in space-time rainfall modeling, geomorphology and scaling in hydrologic processes utilizing a suite of sophisticated mathematical tools, replaced him. She is the current director of the laboratory.

In response to these opportunities, the laboratory started a large program in small hydropower development that built on its previous experience in the development of very large hydropower schemes around the world. The program included a very successful series of professional development courses for hydropower practitioners and developers (Gulliver and Arndt, 1991), major participation in the development of Minnesota's hydropower potential, research on resource assessment, hydroturbine dynamics, aeration and water quality as well as physical modeling and field studies of several small schemes in Minnesota. The laboratory was also in partnership with the City of Minneapolis, the Minnesota Historical Society and Northern States Power Company in the development of a hydropower museum. The need for enhanced research facilities led to the building of the Independent Turbine Test Facility under the sponsorship of the hydropower industry, the US Department of Energy and the LCMR.

At the same time, the US Navy identified the need for a new facility for propeller research with an increased emphasis on hydroacoustics. The laboratory was selected to be responsible for the hydrodynamic design of the Large Cavitation Channel (LCC), which is now the world's largest water tunnel. This was a large project and permitted the laboratory to expand into the field of computational fluid dynamics as a supplement to the extensive experimental research that was required. This led to further work for the German government to develop a similar, albeit smaller (1/2 scale) facility. At the same time the US Office of Naval Research made a significant grant to the laboratory for hydroacoustic research. Faculty from the Departments of Electrical Engineering and Aerospace Engineering and Mechanics collaborated in this research. This latter opportunity fostered much stronger ties with other academic units at the University. In fact, Arndt was appointed Chairman of the Fluid Mechanics Program in 1981, helping to strengthen the laboratory's ties with other fluid mechanics faculty, and foster active research collaboration with Aerospace Engineering and Mechanics, Mechanical Engineering, and Chemical Engineering

A large wind tunnel was designed and built under sponsorship of the NSF. Initially this facility was under-utilized, but now serves as an integral component of a recently developed program in boundary-layer atmospheric turbulence. Support from the LCMR was significant during this period (averaging roughly one-third of the budget for several years). This allowed for significant research in water quality, global warming, engineering solutions to water resources problems, hydrology and a host of other environmental research of critical importance to the State.

In spite of significant growth during the 1980 decade, the laboratory was again under scrutiny by the University. A faculty committee appointed by newly inaugurated President Keller recommended closure along with other major units such as the Colleges of Dentistry and Veterinary Medicine as part of the University's "commitment to focus." This was viewed as a serious threat by all at the laboratory and consumed a good deal of time for the director, faculty, staff and students. A serious erosion of research sponsorship occurred, especially from commercial interests, out of fear that laboratory closure would adversely affect time sensitive projects.

In response to this threat, an external review committee was appointed in 1989. The committee was chaired by Professor Benjamin Liu, Professor of Mechanical Engineering and Director of the Particle Technology Center, University of Minnesota with Norman Brooks, Professor of Environmental and Civil Engineering at the California Institute of Technology, John Cassidy, Chief Engineer, Bechtel Civil and Minerals, Inc., Bob Hansen, Director of the LCMR, Helmer Johnson, Chief, Geotechnical, Hydraulic & Hydrologic Engineering Branch, St. Paul District, US Army Corps of Engineers, Bob Lillestrand, Control Data Corporation and R.S. McGinnis, General Manager, Research, Northern States Power Company. They produced a report that stated, "The current program and long-range goals of the SAFHL are basically very sound in the context of today's funding opportunities and future societal needs. There have been significant evolutionary changes in Lab policy under Dr. Arndt's direction and they should continue."

This report and the receipt of the *Outstanding Water Achievement* Award from the American Water Resources Association greatly encouraged all at the laboratory. However, it took several years to completely recover from the fallout from the proposed closure. Arndt decided at this point to step down (in June 1993) and take a three-year leave of absence to be the director of the Fluid and Particle Processes Program at the National Science Foundation. He returned to SAFL in 1996 to reinvigorate a very active research program in cavitation and hydrodynamics which continues to date. He also provided to the succeeding directors, an invaluable source of information in regard to Laboratory management.

The Directorship of St. Anthony Falls Laboratory remained vacant for the period June 1993 to February, 1995. The Head of the Department of Civil Engineering, Steven Crouch, filled in as Acting Director of the Laboratory during this period. In spite of his many departmental duties, Crouch acted vigorously to familiarize himself with management issues and propose improvements.

In so far as no faculty position was vacant at the Laboratory at the time, it was necessary to seek a new Director from the existing faculty. The clear favorite of all parties was Heinz Stefan, who had served as Associate Director under Roger Arndt. Crouch actively recruited Stefan for the Directorship, but after several months of negotiations Stefan declined the position and the search was unsuccessfully terminated. With no obvious candidate for Director, Crouch named a Steering Committee to manage the Laboratory. The members were Heinz Stefan, Gary Parker and Rick Voigt. Stefan and Parker were (and remain) tenured faculty members, and Voigt was a research fellow managing applied research at the time. Parker served as the committee head. The Steering Committee commenced operation in July of 1993, continuing to February 1995.

The period of governance by the Steering Committee constituted a holding pattern. The Laboratory's infrastructure was aging and its central revolving budget had dipped into the red. There was an element of demoralization among the faculty and staff, who were uncertain about the future of the Laboratory. The Steering Committee attended to maintenance and repairs on an as-needed basis. A method was sought to encourage researchers to redouble their efforts to bring in overhead-bearing projects, so improving the finances of the Laboratory. This was implemented in terms of a devolution of a portion of the overhead earned to the individual principal investigators, to be used as each investigator saw fit.

The budget of the Laboratory improved modestly but perceptibly over time, allowing for a few initiatives. A fund was designated for the purchase of new books for the Lorenz G. Straub Memorial Library at SAFL, with input specifically solicited from the graduate students. An annual "Hostage Exchange" program of invited speakers was organized with SAFL's sister institution, the Iowa Institute for Hydraulic Research. Planning was begun on a Local Area Network, and efforts on minor repairs to infrastructure were redoubled. During this same period, however, an inspection of the electrical wiring, which dated from 1938, revealed numerous code violations.

After several months of negotiation, Steven Crouch appointed Gary Parker to formally take over the Directorship of SAFL from February, 1995, for a term to extend until June, 1998. Crouch is owed a debt of thanks for his services beyond the call of duty in overseeing SAFL in the "interregnal" period between Arndt and Parker. Parker negotiated for and got a "setup package" to allow for a start on an infrastructure upgrade. The package included a) rewiring to bring the Laboratory up to code, b) replacement of the existing inadequate lighting on all experimental floors to fluorescent lighting, c) replacement of the crane and forklift, essential pieces of equipment for a laboratory with a three-dimensional structure, d) repairs to the gate house, e) implementation of the Local Area Network and f) the purchase of "cargo," i.e. a number of pieces of office and experimental equipment. This "setup package" awarded by Francis Kulacki, then Dean of the Institute of Technology, provided a needed lift to Laboratory morale.

Parker commenced the Directorship with the goal of widening the participation of other University of Minnesota Researchers in SAFL. It was during this period that overtures were first made to Vaughan Voller, a faculty member in Civil Engineering, to formally join SAFL. Voller eventually agreed to do so, and his numerical skills remain of great benefit to the Laboratory. Chris Paola, a professor in the Department of Geology and Geophysics, had long maintained a close relationship with SAFL since his appointment in 1983. In the period 1991 – 1993 Paola and Parker worked jointly on an applied project concerning the performance of the tailings basin of the Hibbing Taconite Mine in northern Minnesota. The tailings basin could be modeled as an alluvial fan-delta, a topic in which Paola had a long-standing interest. In the course of the project, Paola proposed the idea of a unique experimental facility, one that could be used to study the morphodynamics of depositional basins, drainage basins and continental margins undergoing tectonic subsidence or uplift. Paola and Parker prepared a proposal to the Academic Research Infrastructure program of the National Science Foundation in order to build this facility, later dubbed the XES (eXperimental EarthScape) facility, a.k.a "Jurassic Tank." Concurrent with Parker's negotiations with Dean Kulacki in regard to the Directorship, Paola and Parker obtained a generous commitment from him to contribute $250,000 in matching funds for the proposal. The proposal was funded, and the XES facility remains one of the mainstays of active research at SAFL. Featured in *Science* (Stokstad, 2000), it is the first system ever built for doing controlled experiments on the formation of large-scale stratigraphic patterns, which develop through the interplay of tectonic subsidence and sedimentation. A plan-view and the mechanism for operation of this basin are illustrated in Figure 6.

In recognition of the growing diversity of laboratory activities beyond hydraulics, the laboratory was renamed during Professor Parker's tenure, from St. Anthony Falls Hydraulic Laboratory to *St. Anthony Falls Laboratory: Engineering, Environmental and Geophysical Fluid Dynamics.* In short, it is simply called St. Anthony Falls Laboratory (SAFL).

In June 1999, Gary Parker stepped down as Director of SAFL, and Efi Foufoula-Georgiou was appointed by Dean Ted Davis as the new Director.

The Laboratory Today (1999-2003)

Efi Foufoula started her Directorship at a lucky time at which a much-needed rejuvenation of SAFL's intellectual power took place. Two new faculty members were added to the Laboratory in 1999 to replace the retired faculty, Professors Charles Song and Cesar Farell. The two new faculty, Miki Hondzo and Fernando Porté-Agel, brought new expertise and new energy that was felt throughout the Lab. Miki Hondzo brought expertise in ecobiological fluid dynamics (the interaction of water with biota), algae formation and biochemical processes in lakes, rivers and the coastal ocean. Fernando Porté-Agel brought expertise in atmospheric boundary layer turbulence, land-atmosphere interactions, field experimentation and large-eddy simulation modeling of boundary layer flows. Both developed very quickly outstanding research programs of national recognition (both received the "CAREER" award given annually to the most promising young scientists by the National Science Foundation). Hondzo developed a new Water Quality Laboratory (also called EcoFluids Laboratory), which features state-of-the-art equipment for biochemical

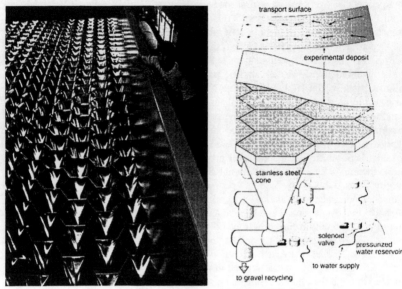

Figure 6. eXperimental EarthScape (XES) facility

analysis. Porté-Agel rejuvenated SAFL's wind-tunnel experimental facility and adapted it for atmospheric boundary layer research.

As the diversity of research expertise within SAFL's faculty saw a sudden growth, Foufoula's first priority was to harness this diversity and promote its uniqueness. Now, faculty could address questions on turbulence, water quality, hydrometeorology, biochemistry, fluvial geomorphology, sedimentology and atmospheric transport, all in the same laboratory and via a unique combination of experimental, computational and theoretical approaches.

Pat Swanson (senior editor, now retired) created the first SAFL web-site (www.safl.umn.edu), which tried to reflect both the uniqueness of each "scientific group" within SAFL, but also the synergy among these groups. The 10 research areas of SAFL are centered on the unifying theme of Environmental and Geophysical Fluid Dynamics. These areas and the group leader in each are:
- Chemical Fate & Transport in the Environment (Gulliver)
- Hydrologic Processes and Multiscale Dynamics (Foufoula-Georgiou)
- Earthscape Processes (Paola)
- Computational Modeling of Transport Processes (Voller)
- Land-Atmosphere Interaction Measurements and Simulations (Porté-Agel)
- Environmental & Biochemical Systems (Hondzo)
- Fluvial & Oceanic Sediment Transport and Morphology (Parker)
- Cavitation & Bubbly Flows (Arndt)
- Lake & River Water Quality Dynamics (Stefan)
- Computational Fluid Dynamics & Applications (Song)

SAFL's mission was revised to read: "Our goal is to advance the knowledge and understanding of environmental hydraulics, turbulence, earthscape evolution and climate, ecosystem dynamics via innovative experimental, theoretical and computational research. As a group, we are committed to transferring that knowledge to the engineering community and to the public through applied research and outreach activities."

SAFL has by no means turned its back on traditional applied-engineering projects. In 1999, a new Associate Director for Applied Research was hired (John Thene) who emphasized quality and cutting-edge technology in the applied projects. Nevertheless, the number of classical model studies the laboratory has performed in recent years has declined for several reasons including a world-wide decline of new hydraulic structures (EOS, 2003), increased use of computational methods in hydraulic design and improvement in the modeling capabilities in developing countries where most major new water projects are taking place. (Many of the present leaders of these projects were trained in labs like SAFL.) SAFL's applied projects within recent years have been very diverse, and have tended to be individually somewhat smaller than in the lab's early years. Some of the more significant ones have dealt with protection measures for bridge piers; drop shaft improvements for the city of San Antonio; visualization of storm-water systems for the city of Milwaukee; and measurement of loss coefficients in pipe fittings for the American Society of Heating, Refrigerating and Air-Conditioning Engineers. We also have a new program in applied research for the oil industry. This involves developing better ways of predicting the geometry of potential oil reservoirs.

SAFL's unique expertise and experimental, numerical and theoretical approaches to water and earth surface processes led to a successful proposal to the National Science Foundation for a new Science and Technology Center named "National Center for Earth-surface Dynamics" (NCED). NCED's (www.nced.umn.edu) mission is "to identify and quantify the major physical, biological and chemical processes that shape the Earth's surface and work towards a holistic approach in which information and understanding from sedimentary geology, geomorphology, engineering, oceanography, hydrology, biology and geochemistry are seamlessly integrated into a consistent, quantitative understanding of Earth-surface dynamics." The hallmark of NCED, as for all Science and Technology Centers, is integrative, multidisciplinary research with clear benefits for society. NCED involves the Departments of Civil Engineering, Geology & Geophysics, and Ecology, Evolution and Behavior at the University of Minnesota, as well as research colleagues from the University of California at Berkeley, Massachusetts Institute of Technology, Princeton University, and Fond du Lac Tribal and Community College. In addition, a strong partnership was formed with the Science Museum of Minnesota on a range of educational programs including a unique series of outdoor, hands-on exhibits on surface processes and engineering. All those who work in laboratories like SAFL understand the fascination of flowing fluids and sediment. The Science Museum partnership will bring some of the fascination of surface dynamics to the public along with the engineering challenges they pose, in a direct and engaging way.

SAFL today is a far cry from the struggling facility that was threatened with closure in the 1980s. Our greatest problem today is insufficient space to house our growing staff and visitors, and the projects they bring. In large part this is because SAFL has continued to develop new ways of applying its core strengths in experimental and theoretical fluid and sediment dynamics. SAFL's success in this regard reflects two mutually beneficial trends. The first is development of a more 'holistic' approach in Civil engineering, in which engineering solutions are more closely integrated with natural conditions and processes. The second trend is a growing appreciation of the role of fluid processes in non-engineering fields such as biology and the Earth sciences. This coincides with an increasingly quantitative, analytical approach in these fields that leads to a greater role for carefully controlled experiments.

SAFL today houses eleven faculty members (a new faculty member not shown in Figure 7 was added this year: Lesley Perg from the Department of Geology and Geophysics who brings expertise in biogeomorphology), approximately 60 graduate students and postdoctoral researchers, 20 undergraduate students and 10 full-time technical staff and research engineers and several support staff. The faculty bring about $1.5 million in research funds annually (mainly from NSF, ONR, NASA, NOAA) and, starting this year, another $2.5 million from the newly established NSF center. Federal funds comprise 85% of the total funds, and the rest comes from state and commercial projects. The revenue created from the overhead of these research grants forms the annual operating budget of the Laboratory. This provides for salary of the support and administrative staff of SAFL and partial support of technical staff and engineers. It also provides for repair and expansion of the lab's experimental facilities and for investing in new initiatives.

The year 2003 has seen major remodeling of the aging building partly due to the new NSF center and partly due to the University's recognition of SAFL as a unique center to promote and cherish in the years to come. SAFL has been completely rewired with new power and digital transmission lines, it has been connected to the main campus with fiber optics, it has been completely re-roofed, and has seen moderate furnishing improvements, especially in the graduate student offices. There is a constant effort to provide the best research and educational experience to our graduate students, and we are grateful to our alumni for supporting the "Graduate Student Fund" created in 1996 for that purpose. More than 400 students have received degrees and numerous undergraduate students have been trained at the Laboratory, and more than 100 visiting scholars and post-doctoral fellows from all over the world have spent up to two years in active collaboration with our faculty and research staff. Also, more than 400 middle and high school students are given tours of the Lab annually to develop an appreciation of water resources and career opportunities. In a typical year, roughly 10 M.S. and Ph.D. students graduate from the laboratory's programs with degrees in Civil, Mechanical and Aerospace Engineering, or in Geology and Geophysics.

The Future

What are the major water challenges in the 21st Century? What science is needed to address these challenges? These two questions have formed the focus of many recent studies by the National Research Council and the National Academies (NRC, 2001a; 2001b). All studies point to the fact that in the face of increasing water demand and other stresses. We have to rethink the traditional approaches to water management and take an integrated view of the water cycle and its interaction with the environment. Water is the main transporting medium for organic carbon and major nutrients that influence terrestrial vegetation processes and aquatic ecosystems at all spatial and temporal scales. These in turn influence the quality, quantity and transport pathways of water, forming a closed cycle of water and bio-ecological interactions. Understanding this cycle holds the key to our ability to make predictions that are essential for efficient water resource planning and management. Theoretical, numerical, and experimental research is needed to achieve this understanding, and SAFL is at the forefront of this research.

One way of understanding the cycle of water bio-ecological interactions is by studying the past. How has the water interacted with the earth surface over the past million years to deposit the sediments and create the variability in landscape we see today? This is an active area of research at SAFL and a component of the "National Center for Earth-Surface Dynamics" at SAFL. The uniqueness of SAFL for such research was recently highlighted in an NSF vision document (NSF, 1999).

"At present there exist only a few major experimental facilities for the study of Earth surface processes. It is important that these facilities be maintained, and that access to them be open to the entire community on a competitive basis...This (St. Anthony Falls) Laboratory has provided through their own work and that of others a significant experimental basis for the understanding of many fluvial and now submarine processes."

We expect that experimental facilities will continue to be central to SAFL's mission, but our focus is on understanding processes, and not on using a particular kind of tool. The facilities will be applied along with numerical modeling, theoretical analysis, and fieldwork. We expect to continue supplementing our "classical" experimental equipment with new facilities such as our Ecofluids Laboratory. The Ecofluids lab also illustrates the increasing role that biological fluid processes will play in SAFL's future. Finally, experiments, field work, and analysis are always most powerful when done in conjunction with theoretical analysis and modeling. Our theoretical research will continue to make use of in-house computing strength, backed by the considerable resources of the University of Minnesota Digital Technology Center and the Minnesota Supercomputing Institute.

References

Arndt, R.E.A. and Marsh, M.H. (1981). *Fluid Mechanics Research in Water Resources Engineering,* St. Anthony Falls Hydraulic Laboratory.

Earth Observing System (EOS). (2003). "Dam removal in the United States: Emerging Needs for Science and Policy." *EOS Transactions*, 84(4), 29.

Gulliver, J.S. and Arndt, R.E.A. (1991). *Hydropower Engineering Handbook,* McGraw-Hill, 1991.

Killen, John M. A. (1974). "Buoyancy-Propelled Test-Body Laboratory Facility." Project Report No. 149, *St. Anthony Falls Hydraulic Laboratory.*

Lamb, H. (1932). *Hydrodynamics,* 6th Ed., Cambridge University Press.

Marsh, M.H. (1987). "The St. Anthony Falls Hydraulic Laboratory, The First Fifty Years." *St. Anthony Falls Hydraulic Laboratory.*

McNown, J.S. (1992). "Recollections of Leaders in Fluid Mechanics." *Bulletin, International Association for Hydraulic Research,* 6:6 - 8.

National Research Council (NRC). (2001a). *Basic research opportunities in Earth Science.* National Academy Press, Washington D.C.

National Research Council (NRC). (2001b). *Envisioning the agenda for water resources research in the twenty-first century.* National Academy Press, Washington D.C.

National Science Foundation (NSF). (1999). "A vision for geomorphology and quaternary science beyond 2000." www.geo.nsf.gov/ear/programs/Geomorph.doc.

Prandtl, L., and Tietjens, O.G. (1934). *Fundamentals of Hydro- and Aeromechanics and Applied Hydro- and Aeromechanics,* 2 volumes, Tans. by L. Rosenhead, McGraw-Hill Book Co. Inc.

Rouse, H. (1976). *Hydraulics in the United States 1776 – 1976.* Iowa Institute of Hydraulic Research, University of Iowa.

St. Anthony Falls Hydraulic Laboratory (SAFL). (1981). *Publications and Motion Pictures.* Circular No. 3.

St. Anthony Falls Hydraulic Laboratory (SAFL). *Research and Facilities at the St. Anthony Falls Hydraulic Laboratory.* Circular No. 5.

Silberman, E. and Ripken, J.F. (1959). *The St. Anthony Falls Hydraulic Laboratory Gravity-Flow Free-Jet Water Tunnel.* Technical Paper No. 24B, St. Anthony Falls Hydraulic Laboratory.

Stokstad, E. (2000). "Seeing a world in grains of sand." *Science,* 287(5460), 1912-1915.

Straub, L.G., Ripken, J.F., and Olson, R.M. (1955). *The Six-Inch Water Tunnel at the St. Anthony Falls Hydraulic Laboratory and Its Experimental Use in Cavitation Design Studies.* Technical Paper No. 16B, St. Anthony Falls Hydraulic Laboratory.

Straub, L.G. and Bowers, C.E. (1956). *The St. Anthony Falls Multi-Purpose Test Channel.* Technical Paper No. 17B, St. Anthony Falls Hydraulic Laboratory.

Straub, L.G., and Anderson, A.G. (1958). "Experiments on Self-Aerated Flow in Open Channels." *Proceedings, ASCE,* 84, HY 7, pp1890-1 to 1890-35.

Index

Page number refers to the first page of paper

Agricola, Georgius 233
Agriculture 207
Albertson, Maury 277
Aqueduct, Dijon 37
Arabia 233
Archimedean screws 233
Austin, Penn. 220

Baumgarten, Andre 24
Bazin, Henry 4, 14, 90
Belanger, Joseph-Baptiste 24, 90
Blaisy tunnel 51
Boussinesq, Joseph 90
Boyce water lift 207
Bridge constrictions 174
Burden Waterwheel 252
Bureau of Mines (USBM) 78

California Institute of Technology 140
Caudemberg, Girard de 51
Cavitation 160
Cermak, Jack 277
Chaillot 24
Chaillot experiments 24, 51
Colorado State University 277
Combes, Charles 24
Cossut, Charles 24
Couplet, Claude Antoine 24
Ctesibius (of Athens) 233

Dams 220; uplift 220
d'Aubuisson, Jean-Francois 24
da Vinci, Leonardo 233
Darcy, Henry 1, 14, 24, 37, 51, 78, 90;
biography 51; birth certificate 1; family 4;
illness 4, 51; letters to Bazin 4; Darcy Medal
51; portrait as youth 4; portrait by F.
Perrodin 51; sitting portrait frontpiece
Darcy's Law 37, 71, 78
Darcy-Weisbach equation 24
Davis, Arthur Powell 220
de Prony, Gaspard-Marie Riche 24, 90
De Re Metallica 233
Dijon 4, 37; water supply 51
Dubuat, Pierre 24

Dust Bowl 263

École Polytechnique 4
Einstein, Albert 140
Einstein, Hans Albert 140
Elephant Butte Dam 220
Europe 233
Experimental EarthScape 289

Fanning, John T. 14, 24
Fayum depression 116
Federal Interagency Sedimentation Project
289
Fick's Law 71
Filtration 37
Flow in Pipes 160
Fluid Transients 160
Fourier, Jean Jacques Batiste 71
Freeman, John R. 160, 220

Garbrecht, Gunther 116
Georgia Institute of Technology 174
Gibbs, Josiah Willard 71
Gilbert, Karl Grove 140
Greece 233
Groundwater 207
Gwinn, W.R. 263

Hetchy Hetch, O'Shaughnessy Dam 220
Historical research methods 116
Hoover (Boulder) Dam 220
Hubbert, M. King 78
Humphrey Gas Pump 252
hydraulic jumps 174

Interagency Sedimentation Committee 140
Iowa Institute of Hydraulic Research 199
Irrigation 207

Kaiser Wilhelm Institute 160
Kindsvater, Carl E. 174

Lane, Emory 277
Lane, E.W. 140, 174
Leer, Blake Van 174

London 51
Lory, Charles 277

Macadamsization 51
McNown, John S. 174
Meade, Elwood 277
Meigs, Montgomery C. 252
Meyer-Peter, Eugene 140
Middle ages 233
Mining 233
Mulholland, William 220

Nagler, Floyd 199
Niagara Falls 252
Non-Darcian Behavior 71
Nutting, P.G. 78

Ohm's Law 71
Onsager, Lars 71
Onsager's Reciprocal Relations 71
Open channels 90

Parshall, Ralph 277
Pergamon water supply 116
Permeability 78
Petit, H.A. 51
Philadelphia water supply 252
Philip, J.R. 71
Philo (of Byzantium) 233
Pipe flow 24
Pipe making 37
Pitot, Henry 14
Pitot tube 14
Pitot-Darcy tube 14
Poncelet, Jean-Victor 24, 90
Porte-Guillaume reservoir 37
Prandtl, L. 14, 160
Pumps 207, 233

Ree, William O. 263
Ree, W.O. 263
Rhine River 140
Rice, C.E. 263
Richardson, Everett 277
Ritter, Charles 24
Rome 233
Rosoir Spring 51
Rouse, Hunter 140, 199, 277
Russell, John Scott 90

Sadd-el-Kafara dam 116
Sediment transport 140
Simons, Daryl 277
Smithsonian Institution 252
St. Anthony Falls Hydraulic Laboratory 289
St. Francis Dam 220
Straub 140
Straub, Lorenz G. 289
Streeter, Victor L. 160
Suzon sewer 37
Swiss Federal Institute of Technology 140

Tennessee Valley Authority 174
Tide mills 252
Trash rack 263
Turbomachinery 160

Unit of permeability 78
University of California, Berkley 140
University of Michigan 160, 199
University of Minnesota 289
Urartu water supply 116
U.S. Army Corps of Engineers 199
U.S. Bureau of Reclamation 277
USDA-ARS Hydraulic Engineering Research Unit 263
U.S. Navy 289
U.S. Reclamation Service 220
U.S. Soil Conservation Service 140

Vanoni, Vito 140
Vegetated channels 263
Velocity distribution 24, 90

Walnut Gulch Flume 263
Water hammer 160
Water supply, Dijon 37; Washington, D.C. 252
Waves 90
Windmill 207
Wyckoff, R.D. 78

Yarnell, D.L. 174, 199